Contents

Notes and Acknowledgments

UNEP has been engaged in developing a long-term programme of work on methodologies for sound environmental management since its inception in 1972. One of the priority areas has been the preparation of an effective EIA, particularly for use by developing countries.

The objective of UNEP's programme has been both promotional and practical. It has sought to identify ways and means through which EIA techniques could be utilized consistently and effectively in development projects and programmes.

It is hoped that decision-makers, in both developed and developing countries, would be persuaded to recognize environmental impact assessments for the important tool they are in promoting sound environmental management. The present guidelines are practical in nature. They are intended to assist in the application of EIA. They attempt through an analysis of present practices to show what problems could arise when an EIA is undertaken and how they could be solved, what are the advantages and limitations of different analytical techniques, and what are the weaknesses and strengths of different approaches (for example, cost-benefit analysis).

The authors are grateful for the many suggestions and comments, both of a procedural and substantive nature, received by them but the responsibility for any remaining deficiencies or errors of fact or judgement rest with them.

An acknowledgement of debt is particularly due to Dr Mostafa K. Tolba, Executive Director of UNEP, without whose personal interest, encouragement and guidance these Guidelines could not have been prepared.

Acknowledgments of debt are also due to Dr Larry W. Canter and to the University of Oklahoma where many of the ideas and practical suggestions contained in the book were initially mooted and discussed.

The successive drafts of the Guidelines were typed by Mrs Jane Maina at UNEP with great accuracy and attention and to her we are very grateful.

Yusuf J. Ahmad   Nairobi
George K. Sammy   October, 1984

HODDER AND STOUGHTON
LONDON SYDNEY AUCKLAND TORONTO

British Library Cataloguing in Publication Data

Ahmad, Yusuf J.
   Guidelines to environmental impact assessments:
   in developing countries.
   1. Environmental impact analysis—Developing countries
   I. Title   II. Sammy, George K.   III. United
   Nations, *Environment Programme*
   333.7′1′091724   TD194.6

ISBN 0 340 38035 7

First published 1985

Photoset by Rowland Phototypesetting Ltd
Bury St Edmunds, Suffolk
Printed in Great Britain for Hodder and Stoughton Educational,
a division of Hodder and Stoughton Ltd,
Mill Road, Dunton Green, Sevenoaks, Kent,
by St Edmundsbury Press, Bury St Edmunds, Suffolk.

Preface

One of the basic premises for sustainable development is the recognition that environment and development are not exclusive of one another but are complementary and interdependent and, in the long run, mutually reinforcing. It is necessary to view environmental problems as a system: a coherent set of solutions is required which will ensure that each step taken, whether in planning or implementation, to meet them fits in perfectly with others envisaged. It is not possible to compartmentalize environmental concerns nor deal with them in sectors. The difficulties are compounded by the fact that we are dealing with a set of mobile and highly dynamic components which makes it necessary that long-term and flexible responses are devised.

It is not surprising, therefore, that while we all speak of sustainable development and agree on its importance and immediacy we have found it extremely difficult to give an operational content to the concept or to identify practical policy guidelines for its realization. Yet, there is increasing and stark evidence that in different regions, notably in Africa, excessive demands are being made on limited resources and the carrying capacity of fragile ecosystems. The unsustainable use, abuse and misuse of the very environmental systems upon which life depends is showing up in soil erosion, lack of water, and in its quality, deforestation, desertification and other adverse natural phenomena to the growing dismay of all of us. This is not to say that we should put a stop to development or to the use of nature to meet our basic needs but that we must do this within acceptable bounds that do not disturb the environmental cycles of life. In so far as renewable resources are concerned, it means reliance on nature's 'income' and not on the depletion of its 'capital'.

One constraint in terms of practical policy is that the causes and effects of environmental problems are complex, inter-locked and, as of now, largely unmeasured. The impacts are frequently synergestic in nature and could sometimes also be irreversible. They are also rather difficult to predict.

We are all aware that important economic values are involved in air and water quality, soil fertility, spread of environmental diseases and so on from the point of view of agricultural productivity, food supply or human health but the problem is to put a price tag on them so as to make them comparable and commensurate with the cost of pollution control measures. And yet we must reach that stage if we are to move away from a definition of development that is concerned solely with changes in national income or capital formation to a more comprehensive and meaningful approach which takes account of the quality of life.

Further refinement of analytical tools and methodologies is necessary which will enable us to achieve an accommodation in economic decision-making of the social and environmental consequences, along with the purely economic ones, of the sometimes irreversible negative impacts of the misuse of our natural capital.

One of the most effective tools we have for this purpose is the environmental impact assessment. Unfortunately, this tool has not been used so far by developing countries to the extent desirable. This has been due to several reasons. The developing countries have been deterred by what they consider to be unduly burdensome intellectual complexities of a multi-disciplinary exercise. Another is the high cost, particularly when outside consultants are utilized.

I believe that neither need to be insurmountable. It is possible to undertake meaningful EIA on the basis of a cost-effective and sim-plified format which would enable the integration of environmental considerations in project and programme formulation. The Guidelines are presented with the hope that they would achieve this goal.

Dr Mostafa K. Tolba
Executive Director
United Nations Environment Programme

Introduction

During the Stockholm Conference and for a number of years afterwards the developing countries needed persuasion that long term and sustainable development could only be achieved through sound environmental management. This is no longer necessary. A number of major ecological disasters (eg the Sudano-Sahelian drought), emerging trends (eg the accelerated rate of urbanization and its impact on the quality of life as in a megalopolis like Mexico City), and visible results of ill-considered and short-term development activities (eg schistosomiasis in the wake of irrigation channels, the destruction of ecological balance of regions due to hydro-electric dams and other constructions) have all tended to drive the lesson home. What has perhaps been most telling is the perceived dependence of development activities in primary producing countries on the exploitation of natural resources (more vulnerable in tropical regions than elsewhere) and the need to preserve them for sustained development. What the developing countries are now demanding, with increasing insistence, is guidance on methodologies, analytical tools, and conceptual frameworks to integrate environmental concerns in development plans, programmes and projects.

It is not only the developing countries but also international development financing institutions that are eager to identify and adapt methodologies to improve project lending, making projects more multi-disciplinary, less sectoral and more responsive to ecological constraints and parameters. To this end, a Declaration of Environmental Policies and Procedures related to Economic Development was prepared by UNEP, the World Bank and UNDP and subsequently signed in 1980, in addition to the three, by the Inter-American Development Bank, Asian Development Bank, African Development Bank, Caribbean Development Bank, Organization of American States, the European Economic Community and Arab Bank for African Development. Since then a committee of the signatories (CIDIE) has been established to review and assist in the implementation of the Declaration. The European Investment Bank signed the Declaration in April 1983.

The Development Assistance Committee of the OECD held a meeting on *Environmental Aspects of Development Aid* in 1982. The meeting concluded that all Members 'should undertake sustained efforts to assure that the operating policies and practises of the development-aid programmes would effectively support their commitment to safeguard the environment'.

II

It is necessary to examine a number of problem areas if environmental concerns are to be effectively integrated in the development process. The developing countries find these problems complex and sometimes obscure, often intractable. The problems tend to be inter-related and of a long-term nature. They require a multi-disciplinary approach. The nature and scope of environmental impact and consequences, especially in the long-term, are speculative. Although recognizing its value, developing countries seldom undertake the preparation of meaningful environmental impact assessment of development projects and programmes. They are deterred by the intellectual complexity (due to multi-disciplinary or adaptive nature) and financial requirements (mounting consultancy fees) involved. Even when the consequences have been identified, it is not always possible to make a quantative (monetary) evaluation of certain environmental effects or the cost of long-term damage. Furthermore, in most developing countries the techniques of integrated physical, socio-economic and environmental planning are not known or practised. Such techniques need to be simplified and made more practical to replace largely sectoral planning. Nor is adequate attention paid to a system of supplementary or satellited accounts that will assign values (if not market, then shadow ones) to environmental goods and services, such as fresh air, clean water, tree cover, soils, genetic resources and the like, so that efficiency in the allocation of scarce resources is improved. In this situation, it is not surprising that for most developing countries the appraisal of development options from the point of view of environmental concerns has remained unattainable.

It is worthwhile to clarify these problems and provide analytical tools that will:
– permit the reconciliation of ecological and economic considerations and values;
– enable environmental considerations to be included in the appraisal of project options at a very early or conceptual stage;
– and thus promote the effective involvement of various concerned groups in development actions.

The fashioning of efficient tools will not only benefit the developing countries. DAC countries of the OECD and their aid policies and procedures will also benefit from the effective integration of environmental concerns in project appraisal and formulation. This comes out clearly in Chapter 10 of the 1982 Review of Development Co-operation which notes that 'the Committee plans to undertake a further review in 1983 of environmental practices of Members' assistance programmes, including consideration at senior policy levels. It was also agreed that evaluation studies will in all relevant cases include assessment of compliance with and the effectiveness of environmental protection provisions of projects. Furthermore, the DAC will also incorporate – where appropriate – environmental considerations into its meetings on sectoral development'.

Amongst these tools one of the most important is the environmental impact assessment (EIA).

III

The primary purpose of this book is not so much to improve the conceptual bases or technological details of EIA statements but the more modest one of trying to determine whether a cost-effective and simplified format for EIA statements with a minimum of financial and other organizational support could be established.

Although it is commonly recognized that environmental impact assessment is a basic tool for undertaking anticipatory-environmental policies and that it has become increasingly urgent to incorporate environmental assessment into the planning and decision-making process from an early stage of the development cycle, developing countries have been reluctant to use the framework of EIA statements systematically. If we take the ESCAP region as an example, we find that formal legislative requirements for EIA statements exist in five countries (Australia, Japan, Republic of Korea, the Philippines, and Thailand), four others have certain requirements of one type or another (for instance, US funded projects require EIA in the Trust Territories of the Pacific Islands, major projects in Papua New Guinea, Malaysia, in specific cases, and guidelines have been developed in India), and the remainder have no formal procedures. This has been due to a number of reasons; environmental impact assessments have proved to be too long, too time consuming in preparation, too expensive (especially when prepared by consulting organizations), sometimes too poorly written and often not as useful to decision-makers as they could be.

Perhaps a more critical reason lies in the fact that EIA statements are prepared by outside agencies. No domestic lobby demands the implementation of the EIA findings – there is no popular recognition of the environmental considerations involved nor training in assessment and evaluation procedures.

The development of cost-effective and practical procedures for EIA, such as we have in mind, would involve consideration of certain basic issues:

Prescription for EIA statements
It is necessary for developing countries to identify and establish the key sectors of development activities that should receive protection because

of substantial and significant environmental impact. In this respect, the important point to bear in mind is that it is neither necessary nor possible for developing countries to undertake the preparation of full-fledged EIA statements for all development activities. In some cases, an informal approximation would suffice. In other cases, extended treatment is necessary. In between, there is a whole spectrum of EIA statements./One way to proceed will be to establish for each country or regional group of countries a list of 4 or 5 subject areas that are basic to their long-term and sustainable development, eg soils in the Sudano-Sahelian countries, deforestation in the Andean region, river-basin development in South-East Asia, etc.\

Scoping of the EIA statements
The scope of the assessment carried out is a function of the funds and technical expertise available. Some of the EIA statements prepared by consulting firms run to several volumes and cost over $3 or 4 million. These types of statements daunt the developing countries and they become reluctant to undertake such costly and cumbersome exercises. The scoping of an EIA statement is, therefore, very important. There appears to be a threshold level of data and information below which an assessment must not fall if it is to be effective for its purpose. But there is a large margin between such a threshold level and what may be considered as the minimum level for effective EIA. The optimal level differs from subject area to subject area and must need to be examined and illustrated in selected cases (eg the cement industry).

Nature of EIA statements
The evaluation must cover in an objective manner both socio-economic and physical impacts. They must also take account of the inter-action between different impacts and their synergistic results.

Awareness-building
The building of awareness in decision-makers of the need to take account of environmental impacts at the conceptual or very early stages of the project cycle is an important goal since the impacts are only likely to impose themselves later as particularly costly exogenous constraints. The need for early assessment has become all the more essential since the EIA process has been increasingly enlarged in recent years to include socio-economic impacts as well.\Unpleasant surprises and unneccessary delays which lead to increased costs could then be avoided. This perception needs to be brought home through concrete examples and data from developing countries' experience.

The inclusion of these elements will enable developing countries to regard EIA statements as the useful and necessary instruments that they are in securing better plans and more efficient (ie less costly) decisions.

The cost of carrying out an environmental impact assessment cannot be considered as high when placed in the perspective of the total cost of a project. In developed countries (eg USA) EIA statements account for approximately 1% of the total cost of projects as compared to 10% normally allocated to planning. But even 1% may perhaps be too high for many developing countries and attempts should be made to rationalize EIA statements and to make them less burdensome and more manageable.

IV

In the reconciliation of environmental needs and concerns with economic goals and constraints EIA can play an important role. But it is necessary to be clear about what it can do and what it cannot do well.

A useful EIA is built upon the prediction of impacts. We have the tools to predict physical, biological and chemical impacts with reasonable accuracy but do not fare so well with socio-economic and cultural impacts. In the event, many significant impacts on social systems are neglected.

Secondly, as a result of past experience, a considerable store of data and information has been built up over the years about the impacts of certain types of projects and programmes, for example, water control schemes. An EIA of such projects is therefore relatively simple and prediction of impact reasonably accurate. Mitigation measures have been identified and viable alternatives developed. This is especially so for large capital projects.

But the current state of environmental degradation, the attrition that population pressure and poverty are exacting from the natural resource base and the regenerative capacity of nature, is not equally susceptible to EIA. Long and medium term remedial measures are possible through improved environmental education and training and anticipatory policies based on the prediction of environmental impacts, but the continuing damage in the short term needs to be managed. Some of the incremental problems, such as tropical deforestation, loss of genetic species and soil erosion are of great urgency and moment and could have irreversible consequences. This type of management requires a move away from sectoral to integrated physical, socio-economic and environmental planning, particularly land use planning and conservation strategies based on a clear understanding of the nature of quasi-option values. It also requires the application of analytical tools, such as CBA, in the day to day trade-offs between present gains and the balance of long term future advantages that the developing countries are called upon to make.

V

Some work, both of a promotional and practical nature, has already been done in the application of cost-benefit analysis to environmental protection but more needs to be undertaken as a matter of priority. Cost-benefit analysis is likely to be a powerful weapon in environmental decision-making once its potentials (as well as limitations) are clearly defined and understood.

In the past, the criticism most often levelled against the application of CBA techniques to environmental protection measures has been that analytical and statistical techniques have not yet produced tools for measuring the more outstanding external dis-economics in an effective and comprehensive manner. In the result, in most CBA calculations, the effects that are easily quantifiable are included prominently and the ones more difficult to quantify are left out so that uncertain, and sometimes unwarranted, conclusions are drawn. This criticism is no longer as valid as it was a few years ago because of recent work done to develop innovative, and sometimes ingenious, techniques to surmount difficulties relating to both the specification and evaluation of environmental impacts. Experimental approaches have also been attempted in which environmental change resulting from development or management alternatives is tested under artificially induced (ie laboratory) conditions. Simulation and mathematical models of alternatives have similarly been used to good effect.

There is, nevertheless, clearly room for fairly simple sensitivity testing to make sure that there is a reasonable or good solution and that great risks are not being taken with the (apparently) preferred options. This requirement underlines the close relationship of cost-benefit analysis and the preparation of EIA statements.

It has become necessary to develop further and refine, through concrete case studies, the analytical tools for evaluation to deal effectively with the following, among other, sensitivity analysis problems:

- risk evaluation in terms of long-term consequences, particularly in respect of events with low probabilities and high or irreversible negative effects (eg desertification, deforestation);
- the allied problem of how to weigh, in comparison with the present, future costs and benefits, ie the problem of discounting and the question of intergenerational ethics;
- non-marginal options or options with non-marginal effects requiring general equilibrium analysis and other techniques;
- situations, as in developing countries, of non-market sectors and steep differences in the distribution of income (thus negating the willingness-to-pay concept).

Yusuf J. Ahmad   Nairobi, September 1984

1
What is EIA?

Introduction

The justification for adding yet another book to the long list of publications that already exist on the subject is to demystify the concept of environmental impact assessment and to present it as a practical and valuable tool for decision-makers in the developing countries. For clarity and convenience, information is divided into a sequence of chapters as follows:
- Chapter 1 addresses the question 'What is EIA?' It presents the basic concepts of EIA, including the fact that EIA is based on predictions;
- Chapter 2 explores several myths which have arisen about EIA: the objective is to clarify these popular misconceptions;
- Chapter 3 proposes a practical framework for EIA, consisting of nine steps;
- Chapter 4 is an acknowledgment of the fact that EIA, as a working tool, has not yet been perfected: it looks at several problems which have been encountered in the past and at some approaches developed to solve these problems;
- Chapter 5 takes a further look at cost-benefit analysis as a tool for environmental decision-making.
- Chapter 6 is a discussion of the institutional arrangements which can facilitate the use of EIA in developing countries;
- Finally, Chapter 7 considers future prospects for the application of EIA in developing countries;

Basic concepts

There exists at present no clear, concise definition of Environmental Impact Assessment. Perhaps it is just as well that this is so. For EIA is still a growing, changing concept, and the lack of a text-book definition facilitates its further development. Having said that, it must also be stated that EIA does *not* attempt to be 'all things to all men'. There is a consensus on several basic tenets of EIA, its aims and its objectives; and these will be presented in this section.

First of all, EIA is a study of the effects of a proposed action on the environment. In this context, 'environment' is taken to include all aspects of the natural and human environment. Therefore, depending on the effects of scale of the proposed action, an EIA may include studies of the weather, flora and fauna, soil erosion, human health, urban migration, or employment, that is to say, of all physical, biological, social, economic and other impacts. Naturally, the number of studies will vary from action to action.

Second, EIA seeks to compare the various alternatives which are available for any project or programme. Each alternative will have economic costs and benefits, as well as environmental impacts, both adverse and beneficial. Naturally, there must be a trade-off between the pluses and the minuses. Adverse environmental impacts may be reduced at higher project cost. Conversely, economic benefits may be enhanced at some environmental cost. EIA seeks to compare all feasible alternatives, and determine which represents an optimum mix of environmental and economic costs and benefits.

Third, EIA is based on predictions. The technical work involved is estimating the changes in environmental quality which may be expected as a result of the proposed action. For example, how will the proposed coal-burning electricity generator affect air quality in the adjacent villages? In the case of some impacts (eg those on water or air quality), prediction can be based on existing mathematical formulae. For others (such as social impacts), numerical analysis cannot be employed. Regardless of how predictions are derived, though, they are not facts and should not be presented as such.

Fourth, EIA attempts to weigh environmental effects on a common basis with economic costs and benefits in the overall project evaluation. If this is done, the decision-maker is less likely inadvertently to overlook an environmental consequence in arriving at his decision. Also he is less susceptible to charges of 'undue influence', which tend to arise when environmental effects are considered separately from economic effects.

Finally, EIA is a decision-making tool. Its ultimate objective is to aid judgemental decision-making by giving the decision-maker a clear picture of the alternatives which were considered, the environmental changes which were predicted, and the trade-offs of advantages and disadvantages for each alternative. The document produced, regardless of its format, should, therefore, include a set of recommendations.

In summary, thus, a 'pseudo-definition' of EIA might be as follows:
– it is a study of the effects of a proposed action on the environment;
– it compares various alternatives by which a desired objective may be realized and seeks to identify the one which represents the best combination of economic and environmental costs and benefits;
– it is based on a prediction of the changes in environmental quality which would result from the proposed action;
– it attempts to weigh environmental effects on a common basis with economic costs and benefits; and
– it is a decision-making tool.

Brief history

During the decades of the 1950s and 1960s, it became increasingly clear that many industrial and development projects were producing unforeseen and undesirable environmental consequences. By the late 1960s, citizen groups had been formed in several countries to address this problem. As a result of the activities of these groups in publicizing the problems of pollution, the words 'ecology' and 'environment' became commonplace in the print and electronic media.

On 1 January 1970, the United States of America had the distinction of becoming the first country in the world to adopt legislation requiring environmental impact assessment on major projects. The National Environmental Policy Act (NEPA) of 1969 was adopted to ensure balanced decision-making.

Following the pioneering effort of the USA, the growth of EIA legislation has been quite phenomenal. Even where legislation is not yet in force, certain governments conduct EIAs on a selective basis. Thus, in terms of actual experience with EIA, more than three-quarters of the developing countries (and practically all industrialized countries) have done impact assessments on at least one project.

It would be well to note there that the pioneering legislation by the USA has not been a universal model for environmental laws. In fact, NEPA has received praise and condemnation in equal measure. What has rightly happened, instead, is that each country has sought to enact legislation which would best fit into its constitutional, economic, social and technological framework. As a result, the present collection of environmental legislation from various countries is a rich and varied mix which accurately reflects a new and growing concept. The selection of an appropriate legislative framework is one of the issues which will be addressed later in this document.

In parallel with the growth of legislation has been the growth of concepts, precepts and techniques of EIA. Within five years of NEPA, a whole library of EIA documents had sprung up. In addition to the Environmental Impact Statements (EISs), there were textbooks, papers and journal articles which proposed a host of methods and methodologies. In the United States alone, there were, by 1976, 26 books and 89 methodologies available to the environmental technologist. So great has been the flow of paper that one eminent US authority on EIA has jestingly proposed an environmental study on the effects of environmental studies! Perhaps his jest should be considered seriously.

In recent years, there has been an awakening of interest in EIA in the developing countries. These Guidelines come in response to that interest. Our objective is *not* to be the headwaters of a river of paperwork such as already exists in the developed world. Instead, we are interested in fostering EIA as a practical tool in the decision-making process. One of the side effects of the 'paper river' in industrial nations has been the partial removal of EIA from the realm of the decision-maker to that of the academic. It would be a great tragedy if this were to repeat itself in the developing nations.

EIA and EIS

Before concluding this chapter on 'What is EIA?', it is desirable to draw a clear distinction between the Environmental Impact Assessment (EIA) and the Environmental Impact Statement (EIS). It is unfortunate that these terms have been used interchangeably by several authors, since they do not represent the same thing.

The terms EIA and EIS both have their genesis in NEPA and the CEQ (Council on Environmental Quality) Regulations which followed it. In the specific context of NEPA and CEQ Regulations, the Environmental Impact Assessment is a brief examination conducted to determine whether or not a project requires an Environmental Impact Statement. CEQ has established a set of guidelines which identify those projects for which a full environmental study would be required. Thus, when a new project is proposed, the EIA is the study of these guidelines. If the project is found to be exempted by the guidelines, a statement of negative findings is filed. If not, work on the full environmental study proceeds and the findings are reported in an EIS. The content and format of the EIS is spelled out in detail in the CEQ Regulations and other documents. Clearly, EIS represents the fundamental activity, and EIA is simply an introduction to it.

The above remarks are specific to the USA. They have been stated here to demonstrate how the confusion of terms come about. In most of the rest of the world, the interpretation of EIA and EIS is far different. Generally, EIA is used to include the technical aspects of the environmental study, including data gathering, prediction of impacts, comparison of alternatives and the framing of recommendations. EIS (if the term is used at all) refers to the document which summarizes the results of the study, and forwards recommendations to the decision-maker. In marked contrast to the US definitions, the EIA in this context is the substantial technical activity, for which the EIS is a necessary reporting device.

It should be evident that much confusion can arise from the out-of-context use of the USA definitions of the terms EIA and EIS. In fact, it has arisen. In this report, the terms will be used in their more widely accepted international context.

2

Exploding myths

Myths flourish in the absence of facts. As a converse, the best way to explode myths is to examine carefully the facts. As with many other new and evolving technologies, EIA has proved a fertile ground for a plethora of misinterpretations and downright falsehoods. It would require a longterm, concerted effort, and far more space than is available in this chapter, to address all the myths about EIA. Such is not our intention. Instead, we have selected a few for attention – those which have in the past served as strong deterrents to the use of EIA in developing countries.

Myth: 'EIA is anti-development'

This statement comes in varying degrees of intensity. One of its mildest versions is 'EIA is just another bureaucratic stumbling-block in the path of development'. A far more radical form is 'EIA is a sinister means by which the industrialized nations intend to keep the developing countries from escaping poverty'. Regardless of the intensity, the root is the same: a perceived dichotomy between EIA and developing activities.

The perception that EIA and development are somehow antagonistic or alternative activities comes from early experience in Western Europe and North America. First of all, the popular movements which preceded environmental legislation were seen as anti-development. In fact, they stood against those developments which were perceived as causes for environmental degradation. Second, the effect of the new laws was to slow some of the development activities, since an EIA was conducted reactively, and this took time. Third, some of the administrators in newly-formed environmental bodies misunderstood their scope of work and, therefore, did work against the expressed wishes of some developers.

If it were true that EIA is anti-development, then the leaders of the less developed nations would probably be justified in rejecting it outright. But EIA is *not* anti-development. It is a tool for development planning much in the same manner as economic analysis. No reasonable person would condemn economic analysis as a planning tool simply because it shows a particular project to be unsound. In fact, when this occurs, the decision-maker is relieved because he has been guided away from a potential disaster.

The same relationship should hold for EIA and development. The objective is to ensure, as far as possible, that potential problems are foreseen and addressed at the appropriate stage of the project design. The role of EIA is sometimes interpreted as an 'approval' or 'disapproval' function. This is only a part of the picture. It is true that an extremely adverse impact which cannot be mitigated will lead to the demise of a project. But such a catastrophic occurrence will probably cause project abandonment in any case. EIA only seeks to ensure that the abandonment comes prior to the commitment of funds for construction, rather than after.

The decisions involved in an EIA are far more complex than a simple 'stop' or 'go' on project progress. Catastrophic side-effects, as just described, are quite rare. More often, the decision is reduced to a question of trade-offs. A series of alternatives are analyzed, each of which has costs and benefits, both economic and environmental. The decision-maker must now answer a question: which alternative best yields the required economic and social benefit at an acceptable financial and environmental cost?

When used to answer the last question, EIA is a complement to development. It has several advantages over the *ad hoc* approach to decision-making that would otherwise be used:
– it is more reliable and less prone to neglect items;
– it is less susceptible to political or personal influence;
– it eliminates the 'reinvention of the wheel' every time a project or programme is assessed; and
– it permits comparison in a systematic, hence reproducible manner.

In summary it can be said that, if used as a tool for balanced decision-making, EIA will enhance the development process in the same way as economic and financial analysis.

Myth: EIA is very expensive'

A genuine concern of many decision-makers in developing countries is the cost of EIA. They fear that, however desirable an EIA may be, they would not be able to afford it. In this section, a look will be taken at what completed EIAs have cost, at the current attitude of funding agencies to EIA, and at workable methods of controlling the cost of EIA.

It must be noted at the outset that the cost of the kind of EIA we have under discussion here is the cost of actually conducting the study and producing the documents. This is the common interpretation of the term. There are more detailed interpretations which include opportunity costs due to delays, cost of research and development, etc, but these are beyond the scope of these Guidelines.

It is unfortunate that so little analysis of the cost of completed EIAs has as yet been done. However, what work has been done shows fair consistency. The cost of environmental studies on waste-water treatment plants in the USA ranged from 0.08% of total project cost for large plants (total cost of more than US$100 million) to 5.4% for small plants (total cost of less than US$2 million). In Thailand, suggested allowances for environmental studies range from 0.1% of estimated construction cost for large projects (construction cost of more than US$250 million) to 1.1% for small projects (construction costs of less than US$1 million). Based on what data is available, a median figure for the cost of EIA would be in the range of 0.5% to 1.0% of project or programme construction cost. This figure is lowest for high-cost programmes, due to the economies of scale.

Regardless of how small the percentage may be, the actual cost of an EIA is not insignificant. A decision-maker who must find US$1 million to finance an EIA is faced with a serious task. If the money cannot be made available, then the environmental study cannot be done.

In the light of the foregoing, it is fortunate that the international lending agencies have increasingly adopted policies which are favourable toward EIA. As a result, it is now proposed in several quarters that the cost of EIA should be included in the overall project funding as a matter of policy. In fact, several multilateral financing agencies already require EIAs of their projects, and pay for them out of project design funds. What is important, therefore, is that the host country's decision-makers should make their EIA requirements known early in the project or programme, and should remain informed as the environmental study progresses.

Once an EIA has been initiated, close limits must be set on its scope. Contrary to certain schools of thought, EIA is not an opportunity for unlimited academic or applied research. Instead, it should seek to provide the best available answers to certain specific questions; and should seek to do so in a cost-effective manner. One approach to achieving this goal is called scoping.

The term scoping is used to indicate a crucial early step in the EIA process. This involves a coarse analysis of the possible impacts of an action with a view to identifying those impacts which are worthy of a detailed study. It is a new development in EIA, and is based on the recognition that the impacts of an action are of varying importance. The idea, then, is first to develop a simple list of all consequences of the proposed action. No attempt is made at this stage to quantify. When the list has been completed, it is carefully examined to identify the important impacts. Naturally, this is region or country-specific. One region or country may consider watershed management very important, but soil erosion unimportant. In another, the reverse order may hold. The point is to control cost and optimize cost-effectiveness by concentrating on the most important impacts.

In summary, the following can be said about the cost of EIA:

- Costs to date have been of the order of 0.5% to 1.0% of overall project or programme cost;
- There has been a growing suggestion that funding agencies should include the cost of EIA as part of project or programme design funding;
- Scoping is a useful step in controlling the cost of EIA and enhancing its cost-effectiveness.

Myth: 'EIA is a paper tiger'

It has been suggested that EIA is an exercise in futility. An action is proposed, designs are made, an environmental report is written to justify the designs, the report is filed away, and the action proceeds as originally envisaged. It is unfortunate that this scenario has been played out many times in both developed and developing countries. In the USA, for example, there has been a great outcry against the use of 'straw-men' alternatives. These are project alternatives which are quite impractical, but which are included simply to make the preferred alternative appear attractive or, at least, palatable.

Another practice which has been deplored in the USA is the reactive EIA. Here, a project is taken through all of its engineering design with no regard for the environmental impacts which may be generated. At the very end, simply to comply with federal regulations, an 'environmental study' is hastily conducted. The objective, though, is not to seek optimal choices, since the relevant choices have already been made. Instead, the objective is to justify the already-completed design.

The EIA which will be advocated in these Guidelines is neither reactive nor concerned with 'straw-men'. It is envisaged as an integral part of the planning process, initiated at project inception. When this is done, real alternatives can be evaluated in a systematic manner, leading to defensible decisions. But for this to happen the decision-maker must make a clear statement of what he requires very early in the project or programme. It is not enough to say 'Bring me what you have and I will tell you if it is enough'. Instead, he must state clearly what data and analysis is required, and must be prepared to make timely decisions when this is presented.

This chapter has considered three of the myths about EIA. It has shown that, with careful planning and monitoring, the EIA process need not be anti-development, nor overly costly, not a 'paper tiger'. With this as background, a step-by-step approach to EIA will be presented in Chapter 3. Each step will be presented and described, and special mention will be made of the timing of the steps and their resource requirements.

pollution of a whole bay would be considered extensive, whereas the pollution of a localized area of the bay would not be so rated.

The significance of an impact looks beyond the magnitude to the actual effects. Consider a species of fish which requires a minimum of 10 parts per million (ppm) of oxygen in the water to survive. If that fish is an endangered species, or if it has economic or recreational value, then a change from 12 ppm to 9 ppm of oxygen, though not great in magnitude, is certainly significant.

The final criterion is region- and country-specific. Different regions of the globe have concerns of environmental sensitivity. In the great cities, it is air pollution. In the Sudano-Sahelian region, it is soil erosion. In South-East Asia, it is river basin management. This criterion simply asks whether any impact of a proposed action will affect an area of special sensitivity.

A first reduction of the list of all impacts is normally made by selecting only those of great magnitude, extent or significance, or which involve areas of environmental sensitivity. If necessary, further reductions can then be made.

The task of reducing the initial list is that of the co-ordinator. In this work he should liaise closely with the decision-maker, and seek assistance from experts or other knowledgeable persons in the fields concerned. However, it should be noted that for this work, home-grown wisdom is often superior to imported expertise. The scoping of the project is best done after the engineering and economic feasibility studies have been completed, when a clear picture of the viable alternatives is available.

Baseline study

The baseline study is simply a record of what existed in an area prior to an action. It is not an end in itself and should not be mistaken for such. Like the Description of the Proposed Action described previously, it is primarily a bench-mark for the future. Thus it need neither be extensive nor all-inclusive.

In course of the scoping exercise, the several most important impacts would have been identified. Since interest would be concentrated on those impacts, it is logical to measure the baseline levels of those environmental parameters which they will affect. Thus, the planning of the baseline survey should flow naturally from the short-list of impacts which is the output of the scoping exercise.

The baseline survey itself will require both field work and review of existing documents. The resources required will therefore be personnel with some basic training in the technical field of interest. In some cases, it will be found that there already exists a person who has devoted years

3

Steps in an EIA

The lack of consensus on the 'best' approach to EIA is freely acknow-
ledged. What is presented in this chapter is one approach to the EIA
process which is believed to be practical, and has the potential for being
cost-effective. The objective is to give the decision-maker a familiarity
with the steps involved, the importance of timing of each step, and the
resources required. This particular approach was developed on the basis
of empirical evidence as collected from developing countries.

The approach to be discussed here consists of nine steps:

1. Preliminary activities
2. Impact identification (scoping)
3. Baseline study
4. Impact evaluation (quantification)
5. Mitigation measures
6. Assessment (comparison of alternatives)
7. Documentation
8. Decision-making
9. Post auditing

In the following sections, each of these will be presented and discussed
in turn. It will be noted that public involvement is not listed as a separate
step. This does not imply that public involvement has been excluded.
Instead, it is a reflection of the fact that the optimum timing and format
of effective public involvement varies from country to country. This will
be discussed in greater detail in Chapter 6.

Preliminary activities

These include a number of first steps that must be taken, including the
identification of questions that must be answered, before an EIA can
start. Amongst these are actions to

- Identify decision-maker(s)
- Select a Co-ordinator
- Decide on work allocation
- Write description of proposed action
- Review existing legislation

Identifying the appropriate decision-maker(s) is much more complex in practical life than may appear at first sight. In many countries (both developed and developing), lines of authority criss-cross and become tangled. Thus, it is very helpful to state clearly which person, or persons, or group, will have the responsibility for making intermediate and final decisions on a project or a programme.

The second preliminary activity is to select a co-ordinator who will manage the environmental study on behalf of the decision-maker. In very exceptional circumstances, the decision-maker can do this management himself. In general, though, this is not the case. A co-ordinator is useful. His mission is to ensure that the study proceeds along the lines set out by the scoping exercise, and that the results generated are in a form that will be useful to the decision-maker.

A third activity is the allocation of work. This can be summarized as the simple question 'Who does what?' There are several alternatives available. In the USA, the developer conducts the assessment and the Environment Protection Agency (EPA) serves in a review and 'watch-dog' capacity. This is appropriate, since the developer *is* the decision-maker, within the constraints of the system. The other end of the spectrum can be found in Bahrain, where an agency of the government actually conducts EIA. One objective here may be to clearly identify those policy decisions which may be made by the government and those other decisions which may be left to the developer.

Between these two models exist a large number of variations. A government may choose to employ an independent consultant (not the engineering design consultant) to conduct the whole EIA. The government may do the non-technical work themselves, and employ a consultant to do the technical calculations. Or they may assume a management role and instruct the developer what technical tasks are to be done. Whatever is decided, it is important that the allocation of work be clearly made early in the life of the project.

Another of the preliminary activities is writing a description of the proposed action. This is a bench-mark statement which will be useful in the scoping exercise and afterwards. It should be brief, certainly no more than ten pages long. It should provide an indication of the problem which the action is intended to solve as well as a list of constraints. But most important, it should clearly specify the proposed action. For example, the action 'increase the potable water supply' is different from the action 'build a dam and water treatment plant'. In the former case, the EIA would probably include alternatives to ground water or desalination. In the latter case, only surface impounded sources would be considered. It is appropriate that the Description of the Proposed Action should be written by the Co-ordinator.

A final acitivity which can be very useful at this stage is a review of all existing laws, regulations and ordinances which would apply to the proposed action. The idea here is to identify areas of possible conflict and avoid them wherever possible. Small items such as transfer of land ownership can balloon into major crises unless they are identified and addressed in a timely manner.

This list of preliminary activities is by no means complete. Instead, it has been limited to those which will be necessary in the majority of projects. Since these are activities which precede the rest of the EIA, it is desirable to complete them as early as possible. Ideally, they should be stated as soon as the project has been identified by the developer or the government. At the very latest, these activities should be done in parallel with the Engineering and Economic Feasibility Studies.

Impact identification (scoping)

The concept of scoping was introduced in Chapter 2, as a means of controlling the extent, and hence the cost, of an EIA. The process usually consists of two parts. First, an exhaustive list of all impacts, severe as well as trivial, is drawn up. Then this list is carefully examined, and a manageable number of important impacts are selected for study. The rest are discarded.

Perhaps the most efficient means of developing a checklist of impacts is by synthesis from other EIAs on similar actions. This synthesis should not be limited to similar actions in one country or region, but should include as many sources as can be obtained. The resources required for developing the list are the co-ordinator and possibly his assistants, and access to completed EIAs on similar actions. One source of information is INFOTERRA (UNEP's referral system). In addition, approaches can be made to national environmental agencies in different countries. Finally, there is a growing number of text-books and source-books which list potential environmental effects of different development or industrial activities.

After developing the checklist comes the task of determining which impacts should be studied in detail. Generally, four criteria should be applied:

- magnitude;
- extent;
- significance; and
- special sensitivity.

Magnitude refers to the quantum of change that will be experienced. A change of great magnitude would be, for example, the doubling of a town's population. In other words, the measured level of the environmental parameter will be twice what it was before.

The extent of an impact refers to the area which will be affected. The to the study of the area of interest. If this is the case, then a baseline study already exists. More generally, though, it must be developed through the study of existing documents and supplemented by field surveys.

It is at the baseline study stage that the technical specialists (if any) make their first major inputs into the EIA. At the end of the scoping step, a list of impacts to be studied would have been generated. Once this is done, the appropriate persons to evaluate these impacts can be identified. It must be strongly emphasized here that the term 'appropriate persons' does not imply 'foreign experts'. In many cases, the needed resource may be available in the host country: a Conservator of Forests, a fisheries officer, an irrigation or construction engineer, etc. Once the specialists have been identified, they should be permitted to guide the activities of the baseline study, so that the data gathered could be used later on to quantify impacts.

Impact evaluation (quantification)

The quantification of impacts is the most difficult technical aspect of an EIA. It is also the most controversial. Perhaps it would be appropriate to deal with controversy first, and the technical aspects later.

It is generally agreed that the quantitative change due to an impact should be computed wherever feasible. It is also agreed that present technology does not permit quantification of all impacts. The thorny question, therefore, is how to treat those changes which cannot be quantified. One approach would be to ignore them altogether, since they represent a considerable level of uncertainty. The other approach would be to include them in a qualitative form. There is an on-going discussion as to which approach is valid, and more will be said about this in Chapter 4.

The problem is compounded when cost is included in the question. The cost of quantification appears to rise geometrically with the degree of accuracy required. What, then, is 'good enough' as far as EIA is concerned?

There is obviously no simple answer to this question. One approach though, could be linked up with the scoping exercise. By looking at the impacts which were ignored, it is possible to get a feel for the 'coarseness' of the EIA. One could then set the degree of accuracy for quantification accordingly. To cite an example: the pollution in an estuary can be estimated by simple formulas, or by complex computerized models. The latter generate far more exact predictions. But if it has been decided to ignore several small, non-point pollution sources, then the degree of accuracy is automatically reduced. In such circumstances, the cost of a computerized model can hardly be justified.

The point to be made here is that judicious scoping should limit both the number of impacts which are studied and the depth to which selected impacts are studied. As previously stated, the quantified impacts are predictions, not facts. Therefore, there is a degree of uncertainty inherent in the process. The objective is to reduce this uncertainty to acceptable levels, not to try to eliminate it altogether. The scoping exercise can and should address the degree of accuracy which represents an acceptable level of uncertainty in light of budgetary constraints.

The resources required for the quantification of impacts are persons competent to do the required calculations or qualitative assessments. These are the technical specialists mentioned in the Baseline Study section. Decision-makers would be well advised to resist the temptation to purchase this expertise wholesale, especially from abroad. A far better approach would be to identify an appropriate individual for each impact, and have this person report directly to the Co-ordinator. In some cases, it may be deemed advisable to have the engineering design consultant work on the quantification of impacts. When this is done, the work of impact evaluation should be clearly separated from the engineering design, so that the function of the Co-ordinator is not circumvented.

As previously noted, it will be the Co-ordinator's role to manage the work of the technical specialists. Specifically, he would be required to ensure that the work of predicting the level of impacts proceeds within the stated scope, budget and time schedule.

The timing of impact prediction (quantification) is bounded by two constraints. First, this work cannot proceed effectively until project alternatives have been defined. Second, it should be finished early enough to permit decisions to be made in a timely fashion. More will be said on this latter constraint in the section on Decision-making.

Mitigation measures

Although it is seldom possible to eliminate an adverse environmental impact altogether, it is often feasible to reduce its intensity. This reduction is referred to as a mitigation measure. Such measures may be engineering works (such as dust collectors, sludge ponds, noise mufflers, etc) or management practices (such as crop rotation, phased plant shut-downs, etc). All mitigation measures have associated costs.

In some respects, mitigation planning is a part of impact evaluation. Once applicable measures have been identified, it is necessary to compute their cost, and to requantify the level of impact, acknowledging the beneficial effect of the mitigation measure. Depending on circumstances, mitigation measures might give rise to two project alternatives where only one existed before. For example, Alternative X may have a given cost and level of pollution. With certain mitigation measures, it may become Alternative M, with higher costs and lower levels of pollution. But the presence of Alternative M does not automatically eliminate Alternative X, and it may be desirable to include both in the final comparison of alternatives.

The same technical specialists who are involved in impact quantification would also work out potential mitigation measures. The timing would essentially be in parallel with the exercise of impact quantification. Again, the role of the Co-ordinator would be to ensure that the work is accomplished within the scope, budget and time schedules established.

Assessment (comparison of alternatives)

The 'Assessment' step has often been labelled 'Comparison of Alternatives'. It is at this point that the technical information gained in previous steps will be pulled together. It is at this point, too, that the environmental losses and gains will be combined with the economic costs and benefits to produce a full picture for each project alternative. The intended output is a series of recommendations from which the decision-maker will choose a course of action.

In order to proceed to compare alternatives, two pieces of information on each project alternative are required. These are:

– a summary of positive and negative environmental impacts; and
– a summary of economic costs and benefits.

The former will have been generated as part of the preceding steps in the EIA. The latter may be developed as part of the EIA, or may come from a parallel economic analysis.

The simplest approach to comparing alternatives across both the economic and the environmental fronts is cost-benefit analysis. To do this, the environmental impacts must be converted into economic equivalents, and listed as costs or benefits. A cost-benefit analysis is then done for each alternative and the recommendations are made on that basis. One attraction of this approach lies in the fact that a large number of decision-makers in the developing countries are quite familiar with economic terms, but ill at ease with environmental concepts. Thus, when the entire project is reduced to a cost-benefit analysis, the decision-maker is being addressed in a language which he understands.

A major problem with cost-benefit analysis for environmental protection is the fact that many impacts cannot easily be reduced to cash equivalents. What is the value of a beautiful sunset, or of uncrowded recreational areas, or of human life? There are procedures for placing economic values on such things, but none are without controversy. Thus, it may be preferable not to set values at all, rather than to set controversial ones.

But if cash equivalent cannot be assigned to environmental impacts, then cost-benefit analysis becomes inappropriate. How, then, can project alternatives be compared on a common basis which includes both economic and environmental inputs? This problem has been recognized in recent years, and a series of solutions have been proposed. These range from a simple ranking of alternatives, to graphical and importance-weighting techniques. Like EIA itself, these methods of comparison are still being evolved. Some of them will be presented in more detail in Chapter 4.

The primary source required in the assessment stage is a human one. Whatever method of comparison is chosen, someone will have to work through the figures and arrive at a preference ranking of alternatives. Ideally, that person should be the environmental co-ordinator. However, if a cost-benefit analysis is performed, assistance from an economist may be needed.

Documentation

The documents which will arise out of an EIA will fall into two categories: reference documents and working documents. The former will contain a detailed record of the work done in the EIA, and are necessary for future reference. The latter are those documents which convey information for immediate action.

Reference documents are intended for use by a technical audience. This audience may include persons working on future EIAs, persons studying the subject project or programme after it has been implemented, or those with some other general interest. Therefore, these documents should be sufficiently detailed to stand on their own. It is here, for example, that charts, graphs and technical calculations would be found.

Reference documents may be a series of reports, each addressing one impact, or they may be one long report containing all the information. Whichever format is used, the contents should be written by the technical specialists who have actually done the quantification of impacts. The Co-ordinator's task would be to ensure that format and presentation is consistent, so that all of the parts can be made into a coherent whole. The writing of reference documents should parallel the impact evaluation stage so that these documents are complete when the comparison of alternatives begins.

Working documents are the formal means of communication from the technologists on the one hand to the decision-maker on the other.

Their primary function, therefore, is to convey clearly information from the former to the latter, so that informed and timely decisions can be taken. This function dictates the format and language of the document. It should be concise and unambiguous. Recommendations should be clearly stated, and reasons for those recommendations presented in summary form.

Since the Co-ordinator's role has been to provide the link between technologists and decision-maker it is logical that he should prepare the working document. This is particularly true if the co-ordinator has himself gone through the mechanics of comparing the alternatives. In any event, the co-ordinator's input will be vital to ensure that the working document which is produced will provide the decision-maker with clear guidance in the final choice of alternatives.

Decision-making

A common assumption in many works on EIA is that the comparison of alternatives constitutes decision-making, and that those making this comparison assume the role of the decision-maker. This is a poor assumption. More often than not, the documentary summary of the EIA is forwarded to a decision-maker who has not been involved on a day-to-day basis with the study. This decision-maker may be one or several government officials, a manager, or a Board of Directors. In any event, decision-making becomes a separate event and should be treated as such.

Within the context of EIA, the decision-making step starts when the working document reaches the decision-maker. In this document will be found a list of project alternatives, with comments on the environmental and economic impacts of each. There will also be recommendations as to one or several preferred courses of action.

It is unlikely that a decision-maker would reject the technologists' recommendations to the extent of selecting an alternative which is clearly labelled 'unacceptable'. However, there will probably be several choices which are generally 'acceptable', and it is among these that the decision-maker must consider political realities along with economic and environmental information. Consider the case of a proposed industry which will discharge a certain volume of liquid waste. One alternative design will result in a level of discharge which is considered acceptable. A second, more expensive design will yield even less pollution. The cost-benefit analysis have put both alternatives on par, so the second has been recommended on environmental grounds. A decision-maker may feel justified in rejecting that recommendation, since he knows that the first, less costly alternative, is far more likely to be built. The point here is that the choice was not between 'bad' and 'good', but rather between 'good' and 'better'. And 'good' was chosen because it was more likely to be realized.

One vital necessity in the decision-making step of the EIA is timeliness. A developer needs to have a decision made in a reasonable time, so that he can know how to proceed. In general, the decision-maker can do one of three things:

– Accept one of the project alternatives;
– Request further study; or
– Reject the proposed action altogether.

If the decision-maker is to accept one of the project alternatives, then the next step would be to complete the engineering designs and proceed with the action. In such a favourable situation, delays are quite unnecessary.

If further study is requested, the decision-maker should be quite specific as to what information is being requested. Non-specific requests for further study can be construed as simply stalling actions, and will reduce the credibility of the EIA procedure. On the other hand, a specific request can usually be easily complied with, thus minimizing delays.

Finally, an outright rejection leaves the project proponent with the choice of filing an appeal or abandoning the proposed action. In order to assist him in deciding on his response, the rejection notice should clearly indicate the grounds for the decision to reject the proposed action.

Extreme delays in the decision-making process will only antagonize developers, and lend credibility to the claim that EIA is anti-development. To avoid this, decision-makers must make every effort to render their verdict in a timely manner.

In summary, it must be remembered that one of the primary objectives of EIA is to aid decision-making. Thus, the working document that is generated must clearly convey to the decision-maker the nature of the problem to be addressed, the alternatives which were considered, the pros and cons of each alternative, and the results of the structured comparison of alternatives. Using this tool, the decision-maker can then make an appropriate choice.

Post audits

When a choice is made, it is assumed that the project or programme will proceed. Is this the end of the EIA? One further step remains to be completed. That step is post auditing.

We have stated earlier that EIAs are based on predictions. Post audits are conducted to determine how close those predictions were to the

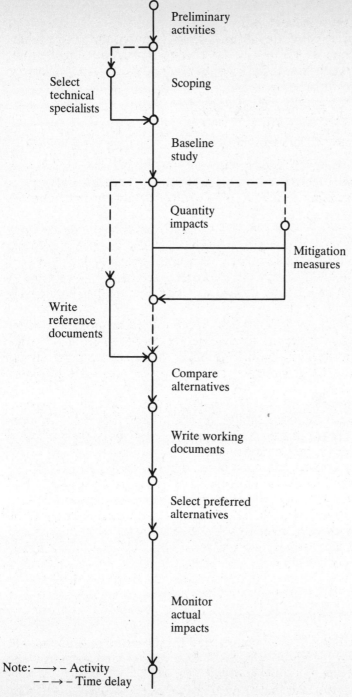

Fig. 3.1 *Activity diagram for EIA*

reality. Such checking forms a valuable data-bank for those who will conduct EIAs in the future.

Because of the duration of the post-audit exercise, it is not normally possible to have it done by the same team which conducted the EIA. Instead, as a terminal activity, the co-ordinator should set up a programme of environmental monitoring, and hand it over to some national agency which collects this sort of data on a routine basis. After a number of years (depending on the nature of the project or programme), the actual changes in environmental quality can be compared with the predicted changes.

Summary

This chapter has presented a step-by-step approach to conducting an EIA. An activity diagram representing this approach is found in Figure 3.1. It is felt that this approach is practical, and has the potential to be cost-effective. There are certain practical problems that may arise in applying this approach. In Chapter 4, several of these problems are identified and possible solutions discussed.

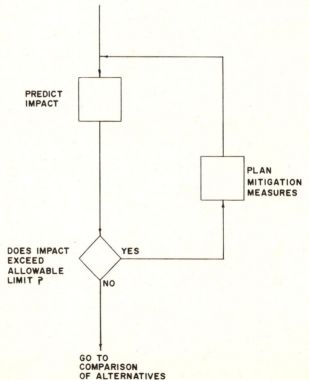

Fig. 3.2 *Relationship between impact prediction and mitigation planning* (see page 15)

4
Problems and Potential Solutions

It was stated early in these Guidelines that EIA is a relatively new and growing technology. As a result, problems are constantly being encountered and solutions sought, both for predicting and mitigating impacts. Often, these solutions are highly innovative and represent a pragmatic approach by those who develop them. This chapter will consider several problems which have been encountered, and give examples of ways in which they have sometimes been solved. The objective is twofold. First, it is an acknowledgment that problems will be encountered. Second, it shows that there are innovative, and often ingenious, techniques to solve the problems.

Problem: too many alternatives

This is a real problem when EIA is introduced at a very early stage, as it should be. Suppose a country identifies the need for mass transport within a city. It will be faced with the choice of the mode of transport (buses, trains, cars, etc); then for each mode there will be choices of type (large or small buses, surface or underground transit, individual or fleet taxis, etc); and finally for each type there will be a choice of routes and of rate structures. The end result could be a bewildering and unmanageable number of alternatives, too large to handle effectively.

Potential solution: the tiered approach
This solution seeks to reduce the problem to manageable proportions by handling it as a series of choices, rather than as one big choice. An excellent example of the successful use of the tiered approach comes from a military application.

The problem here was that the existing training facilities had become crowded. There were a total of 65 individual alternatives available. These included no action, relieve crowding by better management, and acquire more land. Under the last heading fell 21 separate alternative sites and 3 ways to develop each site. It was immediately decided that an across-the-board comparison of 65 alternatives was not feasible and a tiered approach was adopted.

The first tier addressed three basic choices:

1. Do nothing;
2. Relieve crowding by better management; or
3. Acquire more land.

The first tier EIA showed that, while alternative 2 had several immediate advantages (such as low capital cost), alternative 3 was the best long-term choice.

The second tier considered the suitability of 21 alternative sites in terms of cost, availability, accessibility, environmental fragility and downstream effects. From this second-tier analysis it was determined that two of the sites were clearly superior to the others.

The third tier considered the three development strategies at each of the two preferred sites. The objective was to determine the best strategy for each site, but *not* to compare between sites. Thus, this tier consisted of two separate comparisons, one for each site. The output was a best development strategy for each site. As it turned out, these strategies were not the same for both sites.

The fourth and final tier consisted of a straight comparison between Site A, Strategy M, and Site B, Strategy N, where M and N were the best development strategies for sites A and B respectively. A detailed EIA was conducted and a final choice was made.

The use of the tiered approach reduced the duration and cost of comparing a large number of alternatives. Each successive tier involved more detailed studies and inappropriate alternatives were discarded at each level. In this way, the final choice was based on a detailed comparison between two sites using the best development strategy at each site.

The preceding paragraphs describe the actual application of the tiered approach to a real situation. In the hypothetical transport problem presented earlier, a similar approach could be adopted. One set of tiers could be:

Tier 1 – Select Mode of Transport.
Tier 2 – Choose Best Type from Selected Mode.
Tier 3 – Define Most Appropriate Routes for Chosen Type.
Tier 4 – Determine Best Rate Structure for each Route.

In summary, the tiered approach seeks to reduce an unmanageably large number of alternatives by defining the problem in terms of a series of choices. This approach reduces the cost and time of the EIA, while ensuring that all alternatives are considered. The level of detail increases with each tier, so that the final choice is based on an in-depth study of the few most appropriate alternatives. The approach has been successfully applied in several instances.

Problem: too many impacts

Very often, a study of literature on programmes of the type under consideration will reveal that there are hundreds of potential impacts. There simply may not be enough money to study them all.

Potential solution: scoping
The concept of scoping has been presented in detail in Chapters 2 and 3, so only a brief summary will be repeated here. Scoping is a coarse analysis of the possible impacts of a project or programme, with a view to identifying those impacts which are worthy of detailed study. The idea is to optimize the use of available funds by channelling them into a study of the more relevant impacts as against the less relevant. The scoping exercise may also concern itself with the degree of accuracy to which impacts should be quantified. In this case, the idea is to avoid the expense of using highly advanced predictive techniques if in fact such a degree of accuracy is not essential to the judgemental decision-making process.

Problem: lack of data

In both developed and developing countries, it is sometimes found that the historical record of data needed for a proposed action does not exist. The problem then becomes one of finding ways to proceed with the EIA without such data.

Potential solution: synthesis
Synthesis is the process by which data from another location can be used to supplement data at the location in question. Alternatively, it may be the use of one form of data to generate another form of data. Several examples exist and some are presented below.

The first example relates to rainfall and streamflow. When designing a flood control scheme (such as a dam and impoundment) the parameter of interest is streamflow. Unfortunately, streamflow records are comparatively rare. What are readily available in many countries are rainfall records. What must be done, therefore, is to convert rainfall data into streamflow data. Several mathematical formulae have been developed which permit this to be done with fair reliability. Thus, the necessary data can be synthesized, and the engineering design as well as the environmental study can proceed.

A second example of synthesis relates to population growth projections. For the purpose of an EIA, it may be necessary to predict the effect of a new industry, say on the growth of a town. This cannot be done by simply projecting the existing growth pattern, since the new industry will probably cause an influx of job-seekers. One approach which has been used is to study the actual effects of a new industry on the growth patterns of a comparable town and apply similar patterns to the town in question. This is known as a surrogate approach.

The third example again relates to rainfall data. This is the straight transposition of data from one site to another. To do this, a short period of record at the site in question is necessary along with a longer period of record at another site. Such transposition has been done to synthesize total monthly rainfall. The first step is to compare both sets of records for a common period and try to establish a statistically reliable relationship between the two. If this can be done, then the long period of record at the second site can be used to synthesize an equivalent record at the site in question.

The three examples above show how missing records can be synthesized. It is very important to note that synthesis is potentially a very risky exercise, and that it should be approached with great caution. It is particularly important to choose surrogates carefully since the choice of an inappropriate surrogate can lead to totally fallacious predictions. In spite of this caution, situations will arise where the synthesis of data is unavoidable. In such cases, a careful approach can yield sound predictions which will enhance the usefulness of EIA as a decision-making tool.

Problem: lack of expertise

Parallel to the lack of data is the lack of expertise. In many countries, particularly the poorest, there is a shortage of trained technologists and experts to do the work of predicting the changes in environmental quality which would result from a programme or project.

Potential solution: retain management control of the EIA
It is important that, even where the hiring of foreign expertise is inevitable, the host country should retain management control of the EIA. Too often in the past, an EIA has simply been handed over to a firm of foreign consultants and local input has ceased. This is a dangerous error, especially when impacts on the human environment are involved.

A more effective approach is to place the management of the EIA firmly in the hand of a local co-ordinator. This co-ordinator can then make decisions as to what can be done locally and what must be hired from outside. Furthermore, he can exercise close control on the plans and activities of the hired consultants in order to ensure that the work being done is germane to local needs and useful to the decision-maker.

In the context just described, the co-ordinator does not have to be a super-environmentalist, competent in all spheres. In fact, such people do not exist. Instead, the co-ordinator will need

- a general background in environmental work;
- management skills; and
- a good working relationship with the decision-maker.

Even in the least developed countries, such people exist. Some minor up-grading of skills may be deemed desirable, but this can be easily accomplished. The key is to find people who are dedicated and committed to the twin goals of development and environmental betterment.

Problem: impacts cannot be quantified

There are several cases where the theoretical basis for computing the magnitude of an impact does not exist. Thus, there is no available formula or model for calculating the degree to which a proposed action will modify an environmental parameter. Many of these cases pertain to parameters of the human environment, such as migration, culture, etc. This problem was first introduced in the 'Impact quantification' section of Chapter 3. There, the question was asked whether such impacts should be ignored, or whether they should be addressed in a qualitative form.

Potential solution: expert opinion techniques

If an impact has been identified as important during the scoping step, then it should not be ignored simply because its magnitude cannot be quantified. There are several methods which permit the qualitative assessment of an impact based on expert opinions leading to a prediction of its magnitude. Such methods range from simple round-table discussions to the structured Delphi technique.

The first and most difficult step in any expert opinion method is identifying the expert panel. What is sought here is a group whose pooled knowledge represents the state-of-the-art on the question of interest. Care must be taken to ensure that the assembled expertise is real, rather than illusory. 'Years on the job' does not necessarily translate into 'experience', nor 'degrees earned' into 'education'. This is particularly so with social and cultural impacts. A council of tribal elders may have a better feel for the causes of migration than a social welfare officer, for example. Thus, the importance of selecting an appropriate panel of experts cannot be over-emphasized.

When the expert panel has been selected, the next step is the selection of the method of interaction. This, too, must be carefully done to ensure that the best results are obtained. A method which is well-suited to a group of academic researchers may be inappropriate for a group which is more technical in nature. Thus, the expert opinion technique must be fitted to the group of experts involved.

A comparison between the three expert opinion methods is shown in Table 4.1. Meetings and conferences/seminars are face-to-face discussions of an expert group, the former being more flexible than the latter. These methods are subject to several psychological ill-effects: 'noise', dominance and conformity. 'Noise' refers to the many distractions which affect an average committee discussion. For example, during the course of a discussion one opinion may be repeated many times by an individual who champions it. Studies have shown that the statements which are accepted by committees are not necessarily the most relevant or important ones. Instead, they tend to be the ones which have been repeated the most often. Dominance refers to the ability of certain persons to impose their views on others. Thus, the opinion of the most senior member, or even just the loudest member, becomes the consen-

Table 4.1   *Comparison of three expert opinion methods*

	Meeting	Conference or Seminar	Delphi
Effective group size:	Small to medium	Small to large	Small to large
Interaction mode:	Medium	Large	Large
Length of interaction:	Medium to long	Long	Short to medium
Number of interactions:	Varies	Single	Multiple
Format:	Flexible. Could be open or controlled by chairman.	Directed. Presentations follow pre-arranged agenda.	Structured. All interactions go through the monitor.
Costs:	Travel and individuals' time.	Travel, fees and individuals' time.	Clerical, secretarial, individuals' and monitor's time.
Other considerations:	Equal flow of information to and from all. Psychological ill-effects maximized.	Efficient flow of information from few to many.	Equal flow of information to and from all. Psychological ill-effects minimized. Time demands minimized.

sus. Finally, most people tend to conform to peer pressure and avoid radical-sounding positions.

The Delphi technique is an improvement upon the traditional expert opinion approaches to obtaining a consensus of opinion. Like the committee, it draws upon the knowledge of a panel of experts on the subject being investigated. Unlike the committee, the Delphi technique utilizes individual assessment, statistical analysis and controlled feedback to arrive at a consensus. These changes in format reduce the effects of 'noise', dominance and conformity.

When a Delphi study is performed, each panelist is asked to assess the situation independently. The results are then pooled, and statistically analyzed. Each expert is then allowed to study his own responses and the pooled group response. He is then asked to review his own answers in light of the group consensus. These new responses are again pooled and analyzed. If necessary, the process is taken through third and fourth rounds. The advantages are many. Since the same Delphi instrument is circulated to all panelists, the chance of bias due to variations of the questionnaire is removed. The experts work independently, and are therefore not subjected to repetition of arguments or dominance by others. The anonymity resulting from the statistical analysis removes the pressure to conform. Also, the anonymity allows the individual to change his mind without embarrassment.

Expert opinion methods in general, and the Delphi techniques in particular, can be useful tools in predicting impacts upon the cultural environment. By their nature, they are subjective and non-quantitative. However, they can be tailored to give reasonable answers.

Problem: cost-benefit analysis is inapplicable

The 'assessment' step described in Chapter 3 calls for a comparison of alternatives which includes both economic and environmental considerations. It was suggested at that point that, where possible, cash values should be assigned to the environmental impacts; and that the comparison should take the form of a cost-benefit analysis (CBA). It was also pointed out that in certain exceptional situations it will not be possible to assign cash values to environmental impacts. In such situations cost-benefit analysis will not be an applicable instrument for comparing alternatives in an EIA.

Potential solution: numerical methods of comparison
This section contains a selection of the numerical methods which are available for comparing alternatives. When any of these are to be used, they must be preceded by a ranking of alternatives on the basis of economics and a ranking of alternatives on the basis of environmental

impacts. The latter can be achieved by such methods as, for example, the Leopold Matrix, the Battelle EES, or any other matrix or checklist format. The former may be ranked on a straight cost basis or on the basis of cost-benefit analysis. However, if CBA is used, it is very important that environmental costs and benefits be excluded. Otherwise, these will be double-counted: once in the CBA, and again in the comparison of alternatives.

Pairwise comparison This simple analytical tool seeks to simplify the decision-making process by comparing only two alternatives at a time. It is best illustrated by an example.

Consider a programme for which four viable alternatives exist. Economic and environmental studies have been completed, and the costs and benefits of the impacts have been calculated. For some reason, an overall cost-benefit analysis has been deemed inappropriate.

As a first step of pairwise comparison, the group charged with the assessment must familiarize themselves with the results of both the economic and environmental studies. This group may consist of the technologists and economists who have done the studies, the environmental co-ordinator and his staff, the decision-maker or any combination of these.

The group is then asked to compare the first two alternatives and to indicate which is preferable and for what reason. This process is repeated until each alternative has been compared pairwise with every other alternative. The results are recorded in a table as shown in Table 4.2. The scoring is as follows:

1. Compare alternative 'A' with alternative 'B'. 'A' is preferred. Therefore score 1 where row 'A' intersects column 'B' and 0 where row 'B' intersects column 'A'.
2. Compare alternative 'B' with alternative 'D'. Neither is preferred. Therefore score .5 where row 'B' intersects column 'B' and .5 where row 'D' intersects column 'B'.
3. Sum across all rows.
4. The preferred alternative is that receiving the highest score.

Table 4.2   *Example of pairwise comparison*

Compare Alternative	With Alternative				Sum
	A	B	C	D	
A	—	1	0	1	2
B	0	—	0	.5	.5
C	1	1	—	1	3
D	0	.5	0	—	.5

This tool may appear simplistic when only a few alternatives are involved and the preferences are clear-cut. However, its value is seen when a large number of alternatives are present and the preferences are not straight-forward. In such a case, the ability to reduce the problem of choice to a series of pairwise comparisons (even if it is a very long series) is very useful indeed. Its main draw-back lies in the fact that the pairwise comparisons themselves may be considered subjective.

Graphical approach In this approach, the relative economic and environmental desirability of the alternatives are plotted on a graph and the final selection is based on this display. One such graph is shown in Fig. 4.1.

Here, the capital cost of constructing the project is plotted on one axis and the environmental rank on the other. Two points must be noted here. First, rank 1 represents the most preferable alternative. Second,

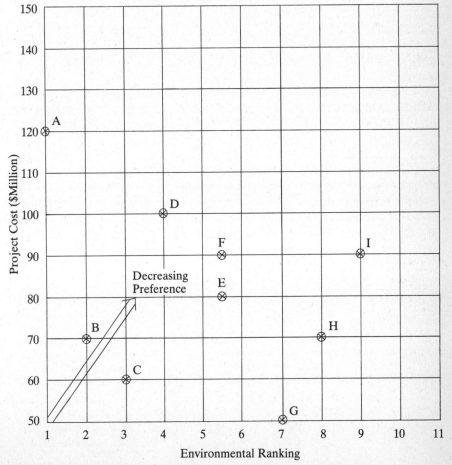

Fig. 4.1   *Graphical approach to comparing alternatives*

where alternatives are equally preferable, they are assigned mid-rank values. Here, E and F are both ranked 5.5 instead of 5 and 6.

The arrow on Fig. 4.1 represents the direction of decreasing preference. In this case, the choice is a toss-up between alternatives B and C. Alternative A, the best choice on environmental grounds, is also the most expensive. Alternative G, the lowest-cost option, is ranked number 7 environmentally. The preferred choices are those which combine low cost with high environmental priority.

A major drawback of this approach is clearly shown in the cited example. That is, no one best choice was identified. Instead, two alternatives were selected as being clearly better than the others. When this occurs, it is necessary to select either economic or environmental considerations as the deciding factor. If the former is chosen, then alternative C would be selected in the example. If the latter is chosen, then alternative B would be selected. Naturally, the use of either economic or environmental considerations as a 'tie-breaker' is a country-specific decision.

Weighted ranking The cornerstone of this approach is the assignment of importance weights to economic and environmental consequences. As before, the alternatives must be ranked in terms of economic and environmental acceptability prior to use of this approach.

Consider the same choices used in the example of the graphical approach. There are nine alternatives and these are ranked in both economic and environmental terms. The rankings are listed on Table 4.3. It has also been determined that the rates or importance of economic to environmental concerns is 7:3. Thus economic concerns are given a weight of 0.7, ie $(7/(7+3))$, and economic concerns are weighted 0.3, ie $(3/(7+3))$. The weighted ranks are the product of the actual rank and the importance weight. These are listed in the third and fifth columns of Table 4.3.

The total score is the sum of the two weighted ranks, and this is

Table 4.3   *Example of weighted ranking*

Alternative	Economic (Weight = 0.7)		Environmental (Weight = 0.3)		Total Score
	Rank	*Weighted Rank*	*Rank*	*Weighted Rank*	
A	9	6.3	1	0.3	6.6
B	3.5	2.45	2	0.6	3.05
C	2	1.4	3	0.9	2.3
D	8	5.6	4	1.2	6.8
E	5	3.5	5.5	1.65	5.15
F	6.5	4.55	5.5	1.65	6.2
G	1	0.7	7	2.1	2.8
H	3.5	2.45	8	2.4	4.85
I	6.5	4.55	9	2.7	7.25

tabulated in the sixth column of Table 4.3. Since the ranking system used awarded rank 1 to the best environmental or economic alternative, rank 2 to the second best, and so on, the overall preferred alternative is the one with the *lowest* total score. In this case, alternative C would be chosen. It is interesting to note that, because economic considerations are weighted so heavily, the lowest-cost alternative moves into the second place overall.

The major challenge of this approach is the assignment of importance weights. These are country-specific and may even be programme specific. It should also be clear that these importance weights can be juggled to favour a particular outcome. Hence, to avoid changes of bias, importance weights should be agreed upon independent of the economic and environmental rankings. One way to ensure this is to agree upon the importance weights before the rankings are done.

The question still unanswered is: How are weights assigned? Essentially, this is the responsibility of the decision-maker. The weights should represent an aggregate view, based on national policy (for government programmes) or company philosophy (for private projects). If the decision-making body has difficulty in coming to a consensus, there are methods of breaking the deadlock. One such method is the Delphi Technique.

This section has addressed the problem of alternatives to cost-benefit analysis for the overall project evaluation. Three numerical methods have been described. Each provides a structured means of making comparisons, but none is without weaknesses. It may therefore be justifiable to use more than one method and compare the results. Finally, it should be noted that this section is not a complete list. There are other methods being developed as well as modifications to those described. The ultimate objective of all of this work is to find a replacement for CBA – which is free from bias – where CBA cannot be applied. This theme is further taken up in Chapter 5.

Summary

The objective of this chapter has been to consider certain problems associated with the conducting of EIA, and to suggest potential solutions to these problems. It is obvious that all problems associated with EIA could not be included. Instead, the discussion has been limited to six of the more common problems:

– Too many alternatives;
– Too many impacts;
– Lack of data;
– Lack of expertise;

– Impacts cannot be quantified; and
– Cost-benefit analysis is inappropriate.

Potential solutions to these problems are (respectively):

– The tiered approach;
– Scoping;
– Synthesis;
– Management approach;
– Expert-opinion methods; and
– Numerical methods of comparison.

There can be no doubt that problems have been encountered and still continue to be encountered in conducting EIAs. As the EIA gains acceptance as a decision-making tool in developing countries, these problems will have to be addressed systematically and solutions sought.

One problem which has not yet been solved is how to evaluate socio-economic impacts (including cultural values, human settlement amenities, basic human needs and the like) at a level of sophistication on par with the evaluation of physical, biological and chemical impacts. It has proved still more intractable to devise tools which would place the different types of impacts on the same grid so that they could be measured in a comparable manner. This problem is not solved by simply getting together a multi-disciplinary team and placing an economist, an engineer, a botanist and a sociologist for example, to work on an EIA. What is necessary is to develop analytical frameworks that will permit assessment procedures to take more explicit account of socio-economic impacts so that at least the more direct and easily predictable impacts on the social system are not overlooked.

5

Cost-Benefit Analysis as a Tool for Environmental Decision-Making

In Chapters 3 and 4, we examined some of the problems that arise when we apply cost-benefit analysis as a component of EIA or when we use it independently (as suggested in the Introduction) to deal with incremental environmental damages in developing countries due to pervasive pressures on the natural resource base and the regenerative capacity of nature. Cost-benefit analysis is such an important instrument for environmental decision-making that a further and more extended discussion has been considered useful and is presented in this chapter.

Quantification and costing of impacts

One of the principal difficulties of cost-benefit analysis relates to the problem of quantification of certain types of environmental damages and, above all, in assigning to them monetary value. Yet it is very important to quantify damage, whenever possible, and to give that damage a monetary value because pollution prevention, control or reduction uses up scarce resources which could be used for other purposes. In deciding upon a scale of pollution control, that is to say whether to use resources for zero or 10 per cent or 30 per cent pollution mitigation, a society is in fact reaching a conclusion as to a good or, the optimal use, of the resources in question. Although it is possible to arrive at such decision-making on the basis of an extended EIA statement or by examining qualitatively the different advantages and disadvantages involved, it can be particularly useful if the benefits, or the avoided damages, of the pollution in question are also evaluated in monetary terms and thus made commensurate and comparable. If this is not done to the extent feasible, then, although it may be possible to carry out a cost-effectiveness analysis of the pollution control costs, it will be difficult to determine the point at which further expenditures on pollution control should cease. It is necessary, however, to guard against undue, unwarranted expectations and hope that quantification will by itself serve to put all issues to rest or that it will lead to the optimal decision. Such expectations are unlikely in most cases to be realized since in any event much quantification is beyond our current reach and depends on value judgements and forecasts.

There are in reality two sets of difficulties. The first set relates to the difficulties of technical specification of environmental impacts. We simply do not know, given the current state of scientific and technical knowledge, all the systemic environmental impacts of activities we undertake. Environmental decision-making is beset with risks and uncertainties about the future and large intangibles. The identification of social and welfare implications of development and investment activities, of change in general and of anti-pollution measures in particular, is still more complex, long-term, and dependent on interdisciplinary effort to predict.

In any event, the availability of adequate hard science data emerges as a critical issue in carrying out cost-benefit analysis. Even when such data are available, it is necessary to determine whether physical functional relationships (eg dose-response functions) can be established in regard to them with a useful degree of reliability and, if they do not exist, whether such functions can be derived. Much of the scepticism about cost-benefit analysis has arisen from attempts to identify or to develop dose-response functions where they simply cannot be constructed at the present time. Needless to say, where such functions do not exist or are subject to errors of large order, it is necessary to proceed with caution: predicate clearly and frankly upper/lower bounds and ask questions such as 'What if?' and 'Is it important?'

It is, indeed, doubtful if any but the most tenuous physical functional relations exist between varying levels of pollution and certain types of environmental damage or amenity benefits, eg loss of wildlife and vegetation, recreational response, damage to cultural assets.

The question arises as to whether in these circumstances it is at all worthwhile, or even useful, to attempt to carry out a cost-benefit analysis. Would it not be better to wait till we have a clearer understanding of the ecological balances and the systemic nature of the environment? The balance of opinion would appear to be that in view of the two-way relationship between the problems relating to the specification of environmental impacts and problems of evaluation, it is worthwhile to learn more about systemic relations through carrying out cost-benefit analysis in different sectors and by employing innovative methodologies for the purpose. It would also appear feasible to overcome some of the specification problems through the use of imaginative statistical techniques recently applied to health and to certain other environmental issues with advantage.

The second set of difficulties relates to evaluation proper, ie the giving of a monetary value to impacts identified.

Willingness to pay as a measure of damage costs

In order to establish such a monetary measure of environmental damage, one obvious and logical way to proceed is to ascertain what people are willing to pay to have the particular damage abated. But it is not always easy or even possible to calculate people's willingness to pay or the 'price' for environmental impacts or the amenity benefits made possible. Apart from the extreme case of placing a monetary value on life itself, there are other difficulties. For instance, when the physical effects of environmental damage are very complex or uncertain, such as those of pesticides, the estimation of a 'price' of the avoided damage becomes unreal. In other cases, it may be extremely difficult to evaluate in practice specific dangers to health and amenity and damage to cultural assets and properties when the impact in question is likely to occur in the future. Apart from a tendency to postpone damage to future generations, and the question of intergenerational ethics thus raised, the synergistic effects of environmental impacts must also be borne in mind.

The application of the willingness-to-pay concept to the fixation of 'price' for environmental impacts is evidently subject to certain difficulties, such as those resulting from the 'free-rider' problem – true individual willingness to pay is unlikely to be revealed because of the expectation that the environmental damage will be abated or treated by someone else. In some cases, an individual may be largely unaware of the environmental dangers to which he is subjected from the existence of one or more pollutions until it is too late. In such cases, willingness to pay on the part of the individual will not determine the price of the environmental protection standards that must be prescribed by society: also 'public policy' value may not equal the individual's value.

A second order of evaluation difficulties arises when, although it is possible to apply the willingness-to-pay concept, the practical means of computing the public's will may be unwieldy or its costs prohibitive. It may be possible to proceed in such cases through the 'revealed preference' route (rise and fall of property values due to differing environmental damage, for instance) but this is dependent on the existence of organized markets and on the need to take account of all factors which affect property values.

Apart from the specification and evaluation difficulties noted above, which are considerable in themselves, cost-benefit analysis as a methodology has a number of technical problems and pitfalls which are aggravated when applied to environmental decision-making. Environmental damage functions are not confined to marginal projects; in fact, the larger the size of the project, the greater the environmental impact and the disturbance to the ecological balance are likely to be. On the other hand, 'marginal' projects may lead to large or non-marginal environmental damage functions – especially if we consider long-term horizons. Evaluation techniques designed for the 'marginal' project may thus suffer from certain inadequacies. Another problem relates to the discount rate to be applied to projects with long-term environmental costs and benefits.

New techniques for assessment

Fortunately, the situation is not as bleak as the enumeration of the difficulties and constraints suggests.

There has been considerable work done of late, both at the theoretical and practical levels, to meet the difficulties that arise in an effective application of cost-benefit analysis to environmental protection measures.

In order to meet the difficulties relating to the specification of environmental impacts, planning approaches have been used that seek to define environmental impacts and establish quantifiable dose-response functions. They can be used either singly or in tandem to deal with specific difficulties. The important thing is to use the right tool or combination of tools, in the right sequence, for the right difficulty. In order to carry out such an operation effectively, it is necessary to know what analytical techniques are currently available, their particular strengths and weaknesses, the nature of the environmental problems that could be treated with them, and the data requirements of each technique.

Cost-benefit analysis cannot be regarded in isolation or as a unique methodology. It is one of a whole range of analytical tools that should be used to obtain effective results. It is necessary, for instance, to begin with the preparation of an adequate environmental impact assessment statement that will specify the effects and consequences of alternative development and management decisions. A cost-effective study will help to establish the lowest implementation cost. Cost-effectiveness analyses are always necessary, even after the most detailed cost-benefit study, since they are important to ensure that the benefits are achieved in the most cost-effective manner possible.

One of the inescapable features of environmental decision-making is the need to face significant uncertainties. Moreover, the environmental impact of certain types of activities is cumulative and becomes noticeable only when they pass a given threshold level at some time in the future. It is, therefore, worthwhile to use risk-benefit analysis as a basic technique of planning. Partial equilibrium analysis cannot always be used for environmental concerns (such as public health programmes) and a general equilibrium analysis in terms of systems analysis becomes worthwhile. Other planning techniques and approaches that could be profitably employed in analyzing environmental impacts include input-output analysis, multiple-objective analysis, optimization models, probability analysis and, in certain cases, trade/investment models. In cost-benefit analysis proper, a number of specific and imaginative techniques have been constructed recently and used with good effect.

Once again, it is necessary to emphasize that these techniques will be useful tools only if the concepts behind them, the data requirements, and their strengths and weaknesses are carefully weighed and assessed in terms of the valuation problems they are expected to solve. They are not techniques of general application although they are directed towards the overall objective of incorporating environmental concerns in management decision-making. As such, they are concerned with issues relating to external dis-economies, the nature of global commons and 'free' or 'collective' goods, the establishment of environmental damage functions, the planning framework for environmental strategies, and so on. These techniques are, however, not to be regarded as mere components of an extended cost-benefit analysis; they can be used independently to resolve specific valuation difficulties.

The definition of damages

It has been noted earlier that valuation techniques can be based on the willingness-to-pay concept.[1] They thus may be classified according to whether they try (a) to establish a demand curve for the benefit in question, (b) to define the opportunity cost of the avoided damage, or (c) to seek recourse to direct estimation of the consumer's preference.

The first type is by far the most useful in gauging, in terms of consumer sovereignty, the nature of possible market preferences, ie the area between the supply and demand curves. Estimation of changes in property values in consequence of different levels of pollution has been widely and successfully used (as well as soundly criticized) in valuing damage from air pollution and noise.[2] It must be admitted, however, that an effective use of this approach requires a number of preconditions, amongst them the existence of a properly organized property market, the absence of 'non-marginal' levels of pollution (eg construction of a new airport). It would appear on balance that greater empirical research would be useful in recreational studies in regard to the further refinement of this approach. Changes in travel costs have also been used as a surrogate market price for pollution, particularly water pollution damage, but more knowledge of the recreational response to changes in water quality is needed. It is easier to calculate the value of damage to livestock, agricultural crops, fruits, vegetables, buildings, etc since market prices for such losses are generally known although often no accurate or substantive estimate of dose-response relationships exists. A great deal of work has been done in establishing linkages between air and water pollution and morbidity and mortality costs, but troubling questions relating to the physical functional relationships and the techniques of valuation remain. Other forms of approach, such as the estimation of present and future damage from pesticides,[3] the survey of wage differentials, the determination of the cost of environmental services, etc require additional research and concrete case studies.

The second type of technique considers the cost of replacing what has been damaged or lost. The specification details, including the nature of the dose-response relationships, are critical here in order to determine adequately or comprehensively the cost of all replacements, value-added costs, alternative costs, etc. One approach used is to consider court awards for damage sustained, but the legal concepts of 'guilt' and 'blame' may affect materially the purely economic valuations of loss and damage. But basically the concept of 'avoided damage' must be regarded as a measure of benefits.

Finally, it is possible to approach the consumers directly and to seek to obtain an indication of their preferences through various types of public opinion polls, market surveys, games of bidding, Delphi techniques, and so on. These approaches can vary from structured interviews to direct questioning. A basic problem arises from the uncertainty which colours a response when the individual is conscious that he is making it not under real but experimental conditions; hence there is often a large measure of probability.

Although the techniques and approaches described need greater time and more empirical testing in terms of concrete case studies, experimental work already carried out provides a number of indications of the nature of environmental costs and benefits. Some of these may briefly be identified, as follows:

- pollution control costs are undoubtedly a burden but often are not a net burden to public authorities or industry because of savings in other costs (pollution damage liability, wasted resources or workers' health) and there is considerable room for savings from built-in control devices;
- the pursuit of unnecessary perfection, ie zero pollution, cannot be justified except in exceptional cases;
- some benefits so far considered as 'intangibles', and therefore non-measurable, can, in fact, be measured using different techniques and approaches and their market values established;
- such measurement throws a different light on the leading benefits of environmental protection measures, at least in the industrial countries, where they are often seen to lie in natural resource – related

recreation and aesthetic experiences generally and in environmentally related health effects; in contrast, beneficial effects on mortality rates are found to have been somewhat exaggerated;
- it is easier to carry out the valuation of the damage costs resulting from air pollution than of other types of pollution because, although the number of receptors is large, the range of the major pollutants and the pollution control techniques are more limited, and, therefore, more susceptible to damage cost calculations;
- it appears possible, with properly defined limits, to apply new economic and related analytical techniques to the valuation of such environmental goods as the preservation of biotic diversity;
- it is necessary to require good reasons before accepting any departures from prices and discount rates prevailing in the market.

In conclusion, a realistic appraisal of the techniques and methodologies that we have considered indicate that, although much useful progress has been achieved in coping with the difficulties and problems of an environmental cost-benefit analysis, much still remains to be done.

In certain areas (identified in the Introduction) cost-benefit conclusions remain very sensitive to the assumptions for key parameters. Even after more extended sensitivity analysis, using 'best assumption' parameter values, it is not possible to provide much useful policy guidance.

These conclusions, however, should not cause undue pessimism. In fact, the present outlook must be considered as promising. Scientific and technological advances are making it increasingly possible to take a long-term and systematic view of the major environmental changes and trends. These advances are expected to accelerate during the coming years. At the same time, it has become possible to deal with many of the evaluation problems through simple sensitivity testing which enables us to discard preconceived preferences, options and alternatives of one sort or another and reach good if not the best solutions.

Notes

1 Accordingly in socialist planning countries and countries with mixed economies, the value of these techniques is lessened and other estimates and decisions are required for the allocation of scarce resources.
2 It is interesting that work on air pollution using this approach has been largely confined to the US, whereas in Western Europe the approach has been used for work on noise abatement.
3 Apart from the direct effects of pesticides causing the poisoning of fish, agricultural products and livestock, as well as humans, secondary and multiplier effects are complex and largely unknown; an extended environmental impact assessment of pesticide use is long overdue.

6

Institutional Arrangements

This chapter looks at the institutional arrangements (the 'back-up' systems) which would facilitate the use of EIA as a decision-making tool in developing countries. The items to be addressed are policy, legislation, personnel, information and public involvement. Of necessity, the presentation will be of a general nature. No attempt will be made to suggest a universally 'best' model of legislation, for example. Instead, the salient issues will be presented so that practical decisions can be taken on a country-specific basis.

Policy

If the twin goals of development and environmental/natural resources conservation are to be achieved, then they must both form part of national policy. Such a policy, clearly documented, will be of benefit to both the developer and the government. It indicates to the former the constraints within which he must operate and it enables the latter to proceed with consistency from project to project.

The first quality of a national policy is that it should be achievable. It serves no purpose to state, as a policy, that no action will be permitted to degrade the environment in any way. This is quite impractical. Instead, any national policy should acknowledge the trade-offs between development needs and environmental quality and set realistic goals for both through sound environmental management.

The second quality of a national policy is that it should be specific. Areas of national concern should be clearly spelled out. On the environmental side, these may be land erosion, air or water pollution, environmental health or natural resources conservation. On the development side, areas of national concern may include employment, housing, industrial development or food production. It is only when these priority areas are clearly identified that the developer (be he private enterprise or a public agency) can plan his actions in harmony with the national programme.

The third quality of a national policy is that it should be flexible. This does not mean that it should be capricious. Instead, it means that policy-makers should be open to suggestion and comment at all times. The stated policy is a chart which is meant to guide the course of development and environmental conservation. It is current at the time that it is published. However, circumstances will change with time and new information will come to hand. The policy must be adaptable in order that it should not become obsolete.

Finally, a policy must be responsive to the social and cultural traditions of the people of the country. Policies which ignore these traditions are soon found to be unenforceable and fall into disuse. Conversely, policies which acknowledge these traditions gain active public support. Such support is vital to the success of environmental programmes. Man is very much a part of the environment and his attitude can dictate the success or failure of efforts to conserve it.

Machinery

A problem facing developing countries in this context is the question of setting up an appropriate environmental machinery. At the time of the Stockholm Conference in 1972 there were approximately 10 countries with some type of environmental machinery. The number has since risen to over 110. These range from full-fledged ministries of environment to one or two officials located in the office of the Prime Minister or the President. The ministries are not necessarily the most effective means of environmental protection – the official located in the Prime Minister's office sometimes wields more power, obtains willing co-operation of different ministers and secures co-ordination of the Departments in order to ensure that environmental concerns are respected in agriculture, industry, human settlements or other areas. Accordingly, it is very difficult to give advice as to what kind of machinery would be useful, but because of the complexity of scientific data and the need for cross-sectoral co-ordination, perhaps it is better to have an environmental machinery with an adequate number of experts who can deal with the different problems which the country faces.

Legislation

As previously stated, there is no universal model for EIA legislation. Instead, there is a varied mix of such laws. At one extreme are those countries which have no laws mandating EIA, but which proceed on a case-by-case basis. At the other extreme are countries which have comprehensive laws, regulations and guidelines covering all aspects of EIA. And, of course, there are many countries in between.

What kind of legislation will be useful? We believe that it will be useful for all countries to have a framework legislation on EIA. At the very least, that will permit the government to insist on an EIA for major proposed actions. In addition it will establish a legally binding set of procedures which seek to safeguard the environment while permitting development.

Many EIA laws include the following:

- a statement indicating when an EIA is necessary;
- an indication of what the EIA must contain;
- a section which empowers a certain body to review the EIA and another body to settle disputes; and
- an indication of the legal/administrative sanctions if the law is not complied with.

When is an EIA necessary?

The statement indicating when an EIA is necessary is of critical importance both to the government and to the developer. It would be unwise to exclude a major programme from the EIA process. But by the same token, it will not be worthwhile to burden a developer with the need for an EIA if his project is a minor one. Thus, the legislation governing EIA should indicate, as clearly as possible, which projects would require EIA and which would not. If it is felt that the requirements would change with time, it may be appropriate to make only a general statement in the body of the legislation and keep the specifics for the supplementary guidelines or regulations. In any case, the rules governing the need for an EIA should be documented.

Four criteria for determining the importance of an impact were presented in Chapter 3. These were:

- magnitude: the quantum of change;
- extent: the area affected;
- significance: the measure of impacts; and
- special sensitivity: country or region-specific concerns.

Generally, these same criteria can be applied to determine if a proposed action warrants an EIA. In addition, two more criteria may also be applied:

- time frame: the expected duration of the effects; and
- irreversability: the commitment of resources such that they cannot be recovered.

What must the EIA contain?

In deciding on format, the end objective should be kept in mind. In Chapter 3, it was suggested that two sets of documents should be produced: those for record purposes, and those for action. The action documents should be reasonably brief, but not at the sacrifice of pertinent detail. At a minimum, they should contain the following:

- a description of the proposed action;
- the results of the scoping exercise;
- a summary of the baseline study;
- a presentation of each alternative considered, including a summary of all expected impacts; and
- the results of the comparison of alternatives, including clear recommendations for action.

The working documents should be clearly cross-referenced to the record document so that it will be easy to locate any detailed information when required.

As noted in Chapter 3, the record documents will be more detailed and more technically-oriented than the working documents. They should be capable of technically justifying any aspect of the EIA and should therefore state all assumptions made. They should also summarize all calculations and list all data sources and other reference materials used. In this context, the record documents should list the names and professional designations of all persons consulted during the course of the study.

Who reviews and who arbitrates?
An essential part of EIA legislation is the establishment of review and arbitration procedures. In cases where decision-making is in the hands of the proponent of the action (ie the developer, be he private enterprise or state agency), then a separate environmental agency may be asked to perform the review function. Their task would be to ensure that the developer's decisions do not run counter to national policies or would not violate areas or concerns of special sensitivity.

If the decision-maker is himself a policy-maker, then the review process may be less formal than that described above. In fact, the review may simply be the study of the working document by the decision-maker-cum-policy-maker. The function of this person or group would then be twofold; to consider the policy implications of the recommendations made and then to make an appropriate decision for action.

Beyond the need for technical review, there is also need for a dispute settlement procedure. It is inevitable that, once decisions are made, objections will occasionally be raised. The objectors may be the developer, any one of the affected parties, or special-interest groups or the general public. It is essential that a body should be nominated to hear and rule on these objections.

Experience has shown that the ordinary courts may not be the best place to refer disputes on environmental impact assessments. In both developing and industrialized countries, courts are overloaded with work and there is some delay in having matters heard. In such circumstances, objectors can effectively use the system as a stalling exercise, even where their arguments hold little merit. Instead, what is required is an independent body which can hear objections and make decisions with reasonable dispatch. Such an arrangement will ensure that EIA continues to be a tool to aid development rather than an impediment to it.

The importance of avoiding delays also holds during the review and decision-making process. Ideally, the EIA should run in parallel with the economic and engineering analysis so that there is no delay in the overall project planning. The review and decision-making step should be accomplished as rapidly as possible. But, to a great extent, this rapidity depends on the quality of the EIA itself. A careful, well-conducted EIA can be reviewed easily. A haphazard, poorly-presented one will always require much time for review. Therefore, the technologist and the co-ordinator who conduct the EIA and prepare the EIS can do much to ensure that review and decision-making run smoothly and quickly.

Sanctions for non-compliance

One of the basic aims of EIA legislation should be to ensure that decision-making is based on balanced, environmentally-sound principles. Therefore, if all proponents of an action adhere to the spirit of the legislation, then the question of sanctions becomes immaterial. Unfortunately, such is not the world in which we live. There will be those who seek to circumvent the intent of EIA and as a result sanctions must be stipulated.

Monetary fines are not always appropriate sanctions in EIA laws. A fine gives the defaulter the option of paying the fine and proceeding with his proposed action. Two other sanctions seem more appropriate here. The first is that of empowering the review agency to halt the proposed action until the necessary EIA has been conducted. Second, the review body may be authorized actually to conduct the EIA themselves, at the developer's cost, in the event of default by the latter. Naturally, each country will need to consider those remedies which are most appropriate in their own national context. But even though sanctions for non-compliance are only a secondary aspect of the law, they must be included if EIA is to be more than just a 'paper tiger'.

Personnel

The lack of local expertise to conduct EIAs in developing countries was highlighted in Chapter 4. A potential solution to the problem, discussed in that chapter, may be the use of a management approach. In this section, that approach will be repeated in summary form.

It has been suggested that in excess of 20,000 trained environmental technologists are required in South-East Asia alone. The figure was derived by employing the traditional approach of summing perceived needs on a sectoral basis. Thus, the total figure quoted would include a number of specialists for water projects, another group for industrial development, a third group for housing programmes and so on. Obviously, this approach is not feasible in the developing countries, at least in the short term.

What is proposed, instead, is a smaller number of generalist managers who can co-ordinate the EIA. Their responsibility would be to under-take the non-technical aspects of the EIA and to manage the work of technical specialists. The technical specialists would be responsible for the baseline study and the quantification of impacts. They will also assist the co-ordinator in comparing alternatives. A major task of the co-ordinator would be to ensure that the technical work is done within budget and schedule, and that the documentary output is in a form that will be useful to the decision-maker.

The proposed management approach is not an end in itself. Instead, it is a short term arrangement to counter the lack of trained personnel in developing countries. As the use of EIA spreads within these countries, it will be necessary to train more technologists who will then assume the responsibility for all aspects of EIA. But in the short term, recognizing the limitations which exist, the management approach is one possible solution to the problem.

Information

The problem of a lack of environmental data in developing countries was also highlighted in Chapter 4. The possible solution which was suggested was the synthesis of data. But a prime requisite for synthesis is the availability of information and case studies on similar projects or programmes in other parts of the world. Therefore, there is the need for a central repository of environmental data.

The central repository referred to above could take one of two forms. In the first place, it could be a library where documents are sent for storage and distribution. In that case, a person seeking information would consult a directory, write to the library and request copies of the documents he has selected.

The second alternative would be for the central repository to simply concern itself with maintaining an up-to-date directory of available documents, but not with storage and distribution of those documents. Here, the interested party would consult the directory to select docu-ments of interest, as well as to find out where these documents are available.

Both the central library approach and the central referral approach described above have benefits as well as drawbacks. Whichever is chosen, there will be the question of staffing as well as funding. UNEP's INFOTERRA system, which maintains a registry of over 800 institutes around the world which deal with environmental impact assessment and close to 10,000 institutes dealing with over 1,000 environmental priority subject areas, is probably the only existing operational international system which comes closest to being an effective central referral system at present. Users from developing countries can only benefit from the extensive INFOTERRA network of experts by seeking a close working relationship with them, as well as with other documentation centres.

In addition to the need for an international environmental data centre, there is also the need for national centres in each country. Such centres would greatly facilitate the flow of information between countries. In the long run, the objective should be an international network of environmental information centres, including (but not limited to) a central repository.

Public involvement

One of the most significant aspects of EIA in many countries is the involvement of the affected public in the environmental impact study. This has taken two forms: the direct involvement of the public, and the inclusion of local values in environmental methodologies. In this section, both of these aspects of public involvement will be discussed, starting with the latter.

The first inclusion of local values in the EIA process comes at the scoping stage. As previously noted, scoping attempts, amongst other things, to differentiate between those impacts which are of importance in the local context and those which are not. By its very nature, the exercise of scoping relies on local perceptions of the environment.

After the scoping exercise, there is a baseline study and quantification of impacts, followed by the comparison of alternatives. Here again, local perceptions of the environment must be included in the computations. If a cost-benefit analysis is performed, then the cash values placed on environmental impacts must reflect the society which will be affected. One society may place great value on an ancient religious site, which another may not. Similarly, one country may be prepared to pay a great sum to preserve the recreational aspects of a river, while another may not. In these examples, and many others, the cost or benefit of an impact is a direct function of the social and cultural values of the people who are affected.

If a cost-benefit analysis is deemed inappropriate, then the alternatives must be ranked in terms of environmental acceptability. One popular technique for doing this involves the use of weighting-ranking or weighting-scaling checklists. A feature of these checklists is the assignment of relative importance weights to different aspects of the environment. As before, these importance weights will vary from one society to the next. There are examples of checklists which have been re-weighted to suit a particular society and the changes from the original weightings are usually significant. An example is the Battelle EES. In its original form, developed for use in the USA, the most heavily weighted block was those parameters related to the physical-chemical environment. When this same checklist was re-weighted to reflect the situation in Thailand, the emphasis shifted to those parameters related to the human environment. Such re-weighting is essential if the comparison of alternatives is to yield meaningful results.

The second form of public involvement is direct comment by the public on the EIA. This is the process which is more generally referred to as 'public involvement' or 'public participation'. The objective is to inform the public of what has been done and to seek their comments. Unlike the inclusion of local values, which is integrated into the technical work of the EIA, public comment usually comes later on, after much of the technical work has been completed.

As with many other aspects of EIA, the optimal forum for receiving public comment will vary from country to country. However, there are some general principles which should be observed. These can be summarized as four simple questions: Who? Why? When? and How?

Who?
Public comment should be sought from all parties who will be affected by the proposed action. Great care should be taken to avoid excluding anyone. Such exclusion, whether intentional or not, will be resented, and those so excluded will often seek redress outside of the structure of the EIA process (eg by demonstrations or lawsuits). It should also be recognized that many individuals will have mixed feelings about a project. One person may favour a project in his capacity as business-man, but object to it in his capacity as sportsman. It is the responsibility of the persons conducting the hearings to try and sort out such mixed feelings in order to arrive at an overall consensus view.

Why?
The rationale behind public involvement is fairly simple. Public projects are implemented to serve a society. It is therefore useful to determine whether the service being provided matches the perceived need. In the case of private projects, the motivation may be personal gain by investors. Here, public involvement is needed to ensure that 'one man's gain' does not become society's 'loss'.

There are six objectives in securing public participation:

- to inform or educate the public;
- to identify problems, needs and values;
- to seek approaches to problem solving;
- to seek reaction (feedback) to proposed solutions;
- to evaluate alternatives; and
- to resolve conflicts.

The public comment which comes close to the end of an EIA will be most useful in attaining the last three of these objectives.

Another utility of public involvement comes out in an indirect manner. Many non-governmental organizations or concerned community groups are increasingly producing state of the environment reports for their region or country. Such reports are invaluable as aids in the scoping process.

When?

In its broadest sense, public involvement is an on-going activity which takes place throughout the EIA. In this sense, the scoping exercise, the work of local specialists, the assigning of cash values and the reweighting of checklists would all be considered public participation.

In the more conventional interpretation, public involvement is the direct comments on the proposed action which are received close to the end of the EIA. There are several reasons for this timing. First of all, the public is better able to grasp and react to clear predictions rather than nebulous concepts. These predictions are simply not available early in the EIA. Second, there is a tendency to lose interest in an issue after a given length of time. Thus, it is possible to get clear reactions and comments over a short period of time, but these become more and more confused if dragged out over a long period of time. It is therefore more advantageous to solicit public comment over a short period towards the end of the EIA, rather than try to sustain public interest over the life of the study. Finally, there is a cost associated with public involvement and this is time-dependant. As a result, it is normally not cost-effective to seek full public comment throughout the life of an EIA.

How?

The most effective means of soliciting public comment varies from country to country. For this reason, the how of public involvement requires careful planning. In the United States of America, Canada and some European countries, public participation has become synonymous with public hearings. In many developing countries, public hearings are the worst possible approach to public involvement. What is required is for the planners of the public involvement exercise to study the objectives we have listed above under the heading 'Why?', and seek to adapt local forums to meet those objectives. It may be found, for example, that informal gatherings yield better results than formal meetings.

In summary, it should be repeated that public involvement is a useful part of any EIA. Each country should therefore seek the most effective ways of eliciting public comment. It is, however, important to note that the interaction between government and people varies from country to country. Therefore, a forum for public participation which works well in one country may fail in another.

This chapter has discussed the institutional arrangements which would facilitate the use of EIA as a decision-making tool in developing countries. It has also concluded the presentation of EIA as a practical tool which can be used to facilitate the development process. In the next chapter, a summary look will be taken at the future prospects for EIA in developing countries.

7

Prospects for the Future

In Chapter 1, it was stated that the purpose of these Guidelines was to demystify the concept and the operational content of EIA and present it as a practical and valuable tool for decision-makers in the developing countries. If this objective has been achieved, it can now be said that EIA is a tool which can enhance the development process by fostering balanced, environmentally-sound decision making. It can also be stated that EIA has definite advantages over the *ad hoc* approach to decision-making and can be cost-effective. What then is needed to ensure its use in the developing countries? Five items have been identified and are presented here for consideration. These are:

- Commitment from leadership;
- Local control of EIA;
- Effective scoping;
- Information sharing; and
- Training.

Commitment from leadership

It is a great credit to the environmental movement that in many countries EIA has grown 'from the bottom up'. However, if EIA is to have its desired impact in the developing countries, this situation will have to change somewhat. Impetus 'from the top down' is desperately needed to augment the popular movements towards environmental awareness. In short, the backing of national leadership is necessary.

The commitment from leadership referred to here is more than simply lip service, or even the passage of legislation. What is required is an awareness of national environmental problems, an insistence upon EIAs where appropriate, and a willingness actually to use the EIA as part of the decision-making process. If such a commitment can be obtained at the level of national leadership, then a country will have crossed one of the biggest hurdles on the road to sound environmental management.

Local control of EIA

If EIA is to be a tool for balanced, environmentally-sound decision-making, then it must be locally controlled. That is to say, the person or group that will use the EIA results should be the ones who dictate its content and direction. Much has already been said about the need for country-specific scoping, the incorporation of local values and the management approach to controlling an EIA. All these issues point to the need for an EIA approach which responds to the needs of the host country, as opposed to the regulations (say) imposed by a funding agency.

Effective scoping

Clearly, the financial and manpower resources available to the developing countries are limited. There is thus a need for practical and cost-effective approaches to EIA. One method of achieving cost-effectiveness is by the use of scoping, that is, identification of the most important impacts for detailed study. This has been discussed in detail in previous chapters.

Effective scoping depends to a great extent on a clear national policy on environment and development. Such a policy would indicate areas of special environmental concern as well as areas of development concern. The former would be invaluable in determining which impacts are important in the national context and which are not. The latter would give some indication of the trade-offs which will permit balanced and sustainable development.

Information sharing

Few, if any, of the world's nations have a sufficiently large environmental data-bank that they would not benefit from exchanges with others. In developing countries, the need for sharing is more than simply desirable; it is imperative. Without it there will be needless duplication of effort and hence duplication of cost.

The approach to information sharing which was suggested in Chapter 6 involves a two-fold thrust. First, each country should develop an environmental data-bank based on its own experiences. And second, there should be a central international repository of environmental data. The general idea would be to form a network involving all national data-banks as well as the central body. This network would facilitate the free exchange of information, with a view to improving the quality of EIAs and reducing costs.

Training

Most of the world's poorest countries are locked in a vicious cycle. Because they have an insufficient number of trained technologists, they are under-developed. But because they are under-developed, they cannot afford to train a sufficient number of technologists. This holds true for many disciplines, particularly the multi-disciplinary character of environmental concerns.

In Chapter 4, it was suggested that one approach to a short term solution might be the training of generalist managers rather than specialists. Such an approach would permit the use of locally-controlled EIA in the immediate future, with specialist input as needed. Once this management system has been put into practice, the developing countries can then focus their attention in the longer-term on training a full complement of technologists, both generalists and specialists, to satisfy their needs.

Statement of objectives

UNEP has set itself the goal of identifying practical, cost-effective procedures for EIA for use in developing countries. This is not an easy task and the end result may be a long time coming. It is, however, an important goal, well worth the effort required for its achievement. What must be recognized, however, is the fact that UNEP cannot achieve this goal by itself. EIA is not a pure science which can be perfected in a laboratory or a 'think-tank'. It is a practical technology which will evolve through actual use in the field. Flexibility is as important as simplicity if the best results are to be achieved. It is vital that developing countries take up the challenge of incorporating EIA into the decision-making process in every way possible. Even in its present form, the technology can be beneficial as a tool for balanced, environmentally-sound decision-making. Through use, lessons will be learnt from successes and failures and ultimately practical, cost-effective procedures for EIA could be established.